Lifestyles of New Zealand forest plants

Text by John Dawson

Photographs by Rob Lucas

Victoria University Press

VICTORIA UNIVERSITY PRESS
Victoria University of Wellington
PO Box 600 Wellington

ISBN 0 86473 205 8

First published 1993

Published with the assistance of
The Lottery Grants Board of New Zealand

Title page: Looking up into the crown of a kahikatea
with nest epiphytes. Lake Pounui, southern Wairarapa.

Printed by GP Print Ltd, Wellington

Contents

Preface

For many millions of years New Zealand's islands have nurtured forest communities containing ancient species that first evolved when dinosaurs roamed the earth. Today, in most temperate areas of the world, many such ancient plants have long been extinct—killed off by Ice Age cold and drought or by competition from more aggressive species that evolved later. In New Zealand many archaic species still survive, protected from the extremes of Ice Age cold by the surrounding oceans and from competition by the isolation of our islands.

From as far back as Cook's early voyage, botanists quickly recognised the distinctive character of the New Zealand forest flora. This generated a keen interest that continues unabated today, particularly amongst botanists and conservationists.

However, the reasons for such enthusiasm may not readily be apparent to the general public. To the casual observer our forests may appear as nothing more than cloaks of drab grey-green vegetation that never change with the seasons, imbuing one with a feeling of monotony. This sombre impression may not be improved, when many of the shrubs and trees produce tiny, inconspicuous white or green flowers. Such plants do not have the immediate appeal of the hordes of exotic plants that now occupy our suburban gardens and public parks.

But, as this book shows, our New Zealand forests have really got it all—for those who are prepared to spend a little time and look. Hidden within and beneath the forest mantle are myriads of plants that can compete on their own terms with any of the exotic imports. New Zealand's forest plants exhibit a seemingly endless diversity of leaf shape and texture, growth forms and specialised lifestyles. Several have colourful flowers too, and many have brightly coloured berries. The abundance of special growth forms in our forests makes them unique in temperate regions of the world.

If our past conservation attitudes are anything to go by, we cannot afford to be complacent about the future of our forest heritage. We hope that this book, by demonstrating the distinctive and fascinating characteristics of our native forests, will take the interested reader beyond the first important stage of recognising species, to a deeper interest in and understanding of their lifestyles and relationships. Such understanding and informed interest is essential if the unique nature and value of our forests is to be widely appreciated and their future assured for the enjoyment of the generations of the twenty-first century.

John Dawson

Rob Lucas

1. Introduction

Most New Zealanders have a special affection for our native forest, or
'bush', as it is inappropriately known. For the Maori this bond with the
world of the god Tane goes back for many generations, but a bond
equally as strong has developed among the descendants of the white
settlers, the Pakeha. Very few of us do not have among our pleasant
memories of childhood the summery sounds of fast flowing water and
the strident din of cicadas, while picnicking by a river in the bush,
shaded perhaps by some of our elegant tree ferns.

For New Zealanders the bush with its frequent abundance of ferns,
nikau palms, vines and perching plants is a familiar, pleasant, but
unremarkable scene. But to the European explorers and later settlers,
with their background of deciduous oaks and beeches and evergreen
pines, the New Zealand forest came as a complete surprise, appearing
to them to be more tropical than temperate in character.

*The scenery along this forest track is, for the whole twelve miles, exceedingly
picturesque. The lofty forest—filled with noble trees of gigantic growth,
clothed not only with their own evergreen foliage, but with innumerable
parasitical plants, ferns, mosses, and orchidae, climbing up to their very
summits—presents a scene of luxuriant vegetation not to be surpassed in the
tropics.*

—Angas, 1847.

*The mode of growth too, the general appearance of a New Zealand forest,
is different from anything in the old world. Thousands of tall columnar
trees of fifty different species, one to two hundred feet high, struggle up
through a wilderness of underwood—their leafy heads so loaded with tufts
of rushy parasites that the true foliage is almost lost in the rank vegetation
of the alien polypiae; whilst innumerable creepers, from the rope-like
supple-jack up to the gigantic Rata (a vegetable boa constrictor) coil round
every stem, run up every limb, glide from head to head, and entwine the
topmost branches of a dozen trees in fifty Gordian knots. The Underwood,
consists of these creepers and of an equally dense growth of young saplings,
mixed with forest shrubs; such as the delicate lady's hair, the Kopakopa, an
elegant plumy fern, the Nikau and many others. And such is the closeness of
the growth, the luxuriance of the vegetation in a New Zealand forest, that
sun and air scarce can penetrate, glimpses only of the sky are caught
through the leafy canopy above, and at high noonday in the fields it is
always green twilight in the woods.*

– –Hursthouse, 1857.

These quotations reflect typical reactions to the New Zealand bush by
early European visitors and settlers.[1,2] * But there were other reactions.
Some found the often tangled growth of such forests overwhelming or
even frightening, or saw it as an obstacle to be overcome in attaining a
more civilised and productive landscape.

* Superscript numbers
refer to the photographs.

So axe and saw felled the larger trees for timber and fire destroyed the rest, leaving initially barren landscapes of blackened stumps and logs—some visible to this day on the hills. The European settlers found about half the country covered by forest. Since then natural forest cover has dropped to about 23 per cent. The blame for deforestation does not all lie with the European settler, however; from their arrival about 1000 years ago the Maori used fire to make clearings for settlements and gardens, to encourage the growth of bracken fern whose rhizomes provided starch, and to keep travel routes clear. Most deforestation occurred when these local fires got out of control, particularly on the drier east of both islands during summer droughts. In this way the estimated 80-85 per cent forest cover at the time of Maori arrival was reduced to 50 per cent by the time Europeans colonised.[3] Evidence for the lowland areas of shrub, fern and grassland at this time having been occupied by forest in pre-human times mostly comes from bits of charcoal and blackened wood in soils, and the pollen, leaves and seeds of trees preserved in swamps. It is estimated that 3000 years ago only about 1.5 per cent of the North Island comprised unforested mountain tops. Due to its higher mountains, the figure for the South Island was about 15 per cent. In addition there were limited areas in the lowlands without trees—sand dunes, riverbeds and swamps.

Clearly, then as now, New Zealand had a forest climate. Extending through 13° of latitude (34–47°S), New Zealand is a narrow land mass with moisture laden oceanic winds coming from all directions. The prevailing wind is north-west and as the axial ranges of both islands lie at right-angles to this, rainfall is heavier on the west than the east. Nevertheless, before the arrival of humans, rainfall was sufficient even in the east to support forest.

The Kinds of Forests

Two major types of native forest can be recognised—conifer broadleaf and southern beech (*Nothofagus*). Both are essentially evergreen, although the former includes a few species such as tree fuchsia and ribbonwood that are leafless in winter.[4]

(a) Conifer Broadleaf Forest[5,6,7]

In the conifer broadleaf forest the conifers provide most of the large trees and mostly belong to the southern hemisphere family Podocarpaceae. Common examples are rimu, kahikatea, totara, miro, matai and tanekaha (celery pine). This family is unusual among conifers in that, although the pollen is formed in proper cones, in most cases the seeds are not. Instead seeds are borne singly at the tips of twigs and either have a fleshy coat[88] or are seated on a fleshy stalk.[84] Other native conifers do have woody seed cones—our largest tree the kauri, also one of the largest trees in the world, and the native cedars, kaikawaka and kawaka.

Broadleaf is a term applied to flowering plants whose leaves are almost always broader than the needle or scale leaves of the conifers. Broadleaf

1. (Opposite above) Watercolour of a forest interior near Papakura, from Hochstetter's book on his travels in New Zealand. Hochstetter visited New Zealand with a German expedition in the 1850s. The colonial government engaged him to explore New Zealand geologically with emphasis on the discovery of useful minerals. It would seem that the artist of this attractive scene, C. Fischer, must have accompanied Hochstetter as most of the plants are fairly accurately depicted—ponga and wheki tree ferns, nikau palm, red and white climbing ratas, the shrub kawakawa and the ground ferns *Asplenium lucidum* and probably *Polystichum richardii*. (Photo: G. Keating)

2. (Opposite below) Von Tempsky was a Prussian mercenary and adventurer who was involved in the wars against the Maori. At quiet moments he would record the scene in watercolours. Although his style could be termed 'primitive' he had a good eye for the more striking forest plants. In this scene he and his men (he is at far left) relax or cook by a stream. Toetoe, a tree fern, cabbage tree, nikau palm, long-leaved kiekie vine and other looping vine stems capture the profusion of the 'bush'. (Photo: G. Keating)

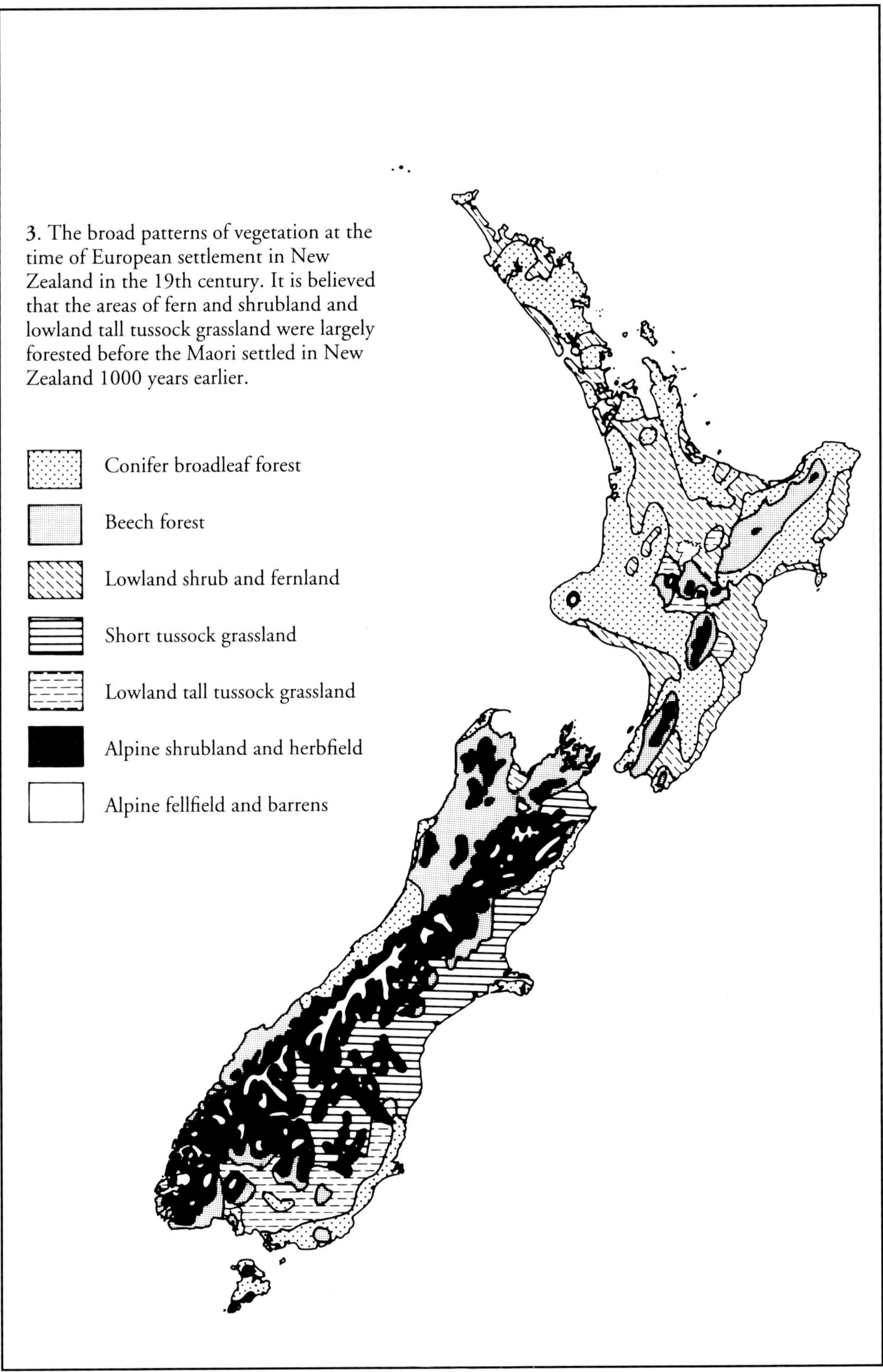

3. The broad patterns of vegetation at the time of European settlement in New Zealand in the 19th century. It is believed that the areas of fern and shrubland and lowland tall tussock grassland were largely forested before the Maori settled in New Zealand 1000 years earlier.

species, ranging from tall trees, such as rewarewa, northern rata and pukatea, through smaller trees to shrubs, vines and herbs, greatly outnumber the conifers.

Conifer and broadleaf plants reproduce by multicellular seeds and are considered to be highly evolved. But these forests contain more primitive plants too that reproduce by single-celled spores. Notable here are the ferns, including several handsome species of tree fern, mosses, liverworts, lichens and fungi. All are abundant.

The most striking features, though, of the conifer broadleaf forest are those it shares with tropical rain forest. There is the same complex stratification with scattered tall emergent trees, mostly conifers, over a canopy or forest roof of largely broadleaf trees. Below this come subcanopy trees, small trees and shrubs and finally the forest floor where ferns predominate. But the most eye-catching plants are those with specialised growth habits, the many vines—from slender ferns to several broadleaf species with massive woody cable-like stems—and epiphytes or 'perching plants'. Some epiphytes form massive 'nests' on tree branches, while others, such as ferns and orchids, hang below them; still others are shrubs or even trees which eventually send massive roots to the ground.

Our conifer broadleaf forests are most similar to certain montane tropical forests in New Guinea. They differ most notably from lowland tropical forests in having fewer species, the presence of conifers, fewer palms (only one species, the nikau), more abundant tree ferns, and smaller leaf sizes.

4. Winter time in the Akatarawas near Wellington. The tree fuchsia (*F. excorticata*) with its leafless branches stands out among the evergreens. To the right the semi-deciduous wineberry (*Aristotelia serrata*) has a scattering of yellow leaves.

5. (Above) Exterior view of conifer broadleaf forest, near Whangamomona, inland Taranaki. Tree ferns occupy the gullies. The main canopy of the forest is very uneven and emergent trees stand out on the skyline. (Photo: John Dawson)

6. (Opposite above) Conifer broadleaf forest in Westland with rimu emergent over a canopy dominated by kamahi and southern rata. Mt Cook in the background, New Zealand's highest mountain.

7. (Opposite below) Interior view of a conifer broadleaf forest at the Wainui Waterworks Reserve near Wellington. The large trunks are of emergent rimu and northern rata, the more slender trunks are mostly miro. Foliage of the main canopy tree kamahi appears at top right. A shrub layer includes the ponga tree fern. Two different nest epiphytes can be seen: at upper centre the close, yellowish-green tufts of *Collospermum hastatum* and at upper left the more grassy *Astelia solandri*. One of the miro trunks in the background is closely covered with the foliage of the white climbing rata.

Species composition alters with rainfall and altitude. In wetter western and central areas the commonest conifer emergent is rimu in association with several other conifers, and there is a wide range of broadleaf trees and an abundance of tree ferns. In forest at higher altitudes, where beech forest is absent, mountain totara and mountain cedar (kaikawaka) may be emergents over the broadleaf kamahi and/or a number of other smaller broadleaf trees. In the drier east of both islands rimu is absent from most forests, although other conifers—kahikatea (swampy sites), matai and totara—are often common. The broadleaf trees are mostly smaller species and tree ferns are few. The South Island inland basins are even drier and they were occupied by low forests dominated by mountain totara, mountain celery pine and small broadleaf species.

Forests dominated by the kauri are mostly restricted to the Coromandel and Northland peninsulas of the northern North Island, and there mostly to infertile sites.

Where drainage is impeded, but stream fed, relatively fertile swamps eventually support tall kahikatea forest, often in association with pukatea and swamp maire. Forests on largely rain fed and consequently infertile bogs, particularly in the west and south of the South Island, are dominated by the podocarps—silver pine or yellow silver pine.

Species diversity decreases from north to south and species tend to reach their southern limits in groups. At about 38°S in the north-central North Island, such notable trees as the kauri, pohutukawa, puriri, mangeao, taraire and tanekaha drop out, although the latter reappears in the northern South Island. In the northern South Island at about 42°S, tawa, kohekohe, rewarewa, northern rata and the maires, among others, reach their southern limits. A number of specialised vines and epiphytes also go no further south than this region.

(b) Beech Forest (*Nothofagus*)[8,9,10,11]

Beech forest as it is commonly known is a simpler forest with fewer species than conifer broadleaf forest. Typically there are no emergent trees. The canopy is usually formed exclusively by species of *Nothofagus*, and vines and larger epiphytes are rare or absent. In regions of low to moderate rainfall, the forest is fairly open underneath with some small-leaved shrubs and ferns and mosses, but there are also open areas of forest floor with deep carpets of fallen beech leaves. In wetter western localities there is a greater abundance of mosses, lichens and ferns, both on the ground and on the tree trunks and branches, and a greater variety of shrubs and small trees.

Collectively the beech species are more tolerant than most conifer broadleaf species of cold, drought and infertility, so the usual pattern is for conifer broadleaf forest to give way to beech forest at higher altitudes and latitudes and also on the thin soils of ridge crests in the lowlands.

The beech species all have small leaves, some with teeth and others without. Silver beech leaves last several seasons, but in the others they

8. (Opposite) Exterior view of silver beech forest near Lake Manapouri showing uniformity in colour and texture.

last one year, falling soon after the new leaves expand in the spring. Mistletoe parasites may be conspicuous in beech forests, with yellow and red-flowered[12,173] species providing splashes of colour against the dark foliage. A striking feature of the branches of some silver beeches is the parasitic yellow-orange golfball-like fungus *Cyttaria*.[13]

Beech forests are absent from some regions where they might be expected, notably Mt Taranaki, central Westland and Stewart Island. Perhaps in its slow expansion over recent millenia beech forest has yet to reach these places. Stewart Island though may be unattainable for the beeches as their seeds travel only short distances.

Since the last glaciation fossil evidence indicates that conifer broadleaf forest, dominated by matai, spread first through the North Island. There was little evidence of beech forest. From about 10,000-7000 years ago was the warmest and wettest phase of the postglacial, and conifer broadleaf forest, dominated by rimu in the west and matai and kahikatea in the east, was wide spread through both islands. Beech forest was still very limited. Since then there has been a trend towards cooler and more variable climates, and as a consequence, beech forest has expanded at the expense of conifer broadleaf forest. This process still continues and may perhaps explain forests intermediate between the two main types to the north and south of the South Island 'beech gap' and at intermediate altitudes on some mountains.

The Ancient History of New Zealand Forests

The evidence of spore, pollen, leaf and sometimes seed fossils in sedimentary rocks shows that conifer broadleaf and southern beech forests have been present in New Zealand for a very long time, perhaps as long as 100 million years.

For most of that time climates were mild to very warm and moist, and landscapes were gentle. It wasn't until recent geological times that mountain building, and the Ice Age with its alternations of glacials and warmer interglacials, greatly affected the forests. A number of species became extinct, including several large-leaved species of southern beech related to a group now restricted to New Caledonia and New Guinea, and several species of palm including a small-fruited coconut. It is suggested that during each glacial, forest was restricted to the north of the North Island and expanded southwards during the interglacials. Our oceanic climate might also have permitted the survival of pockets of forest in southern parts where there was shelter and a northerly aspect. With the return of interglacial warmth these pockets would expand to reinstate the forest cover.

9. (Opposite) Lake Te Anau. In the foreground low mountain beech at treeline. Across the lake red and silver beech forest below changing to smoother mountain beech forest above.

10. (Above) Interior of a hard beech forest on a ridge, Kaitoke Regional Park, Upper Hutt. Species are few and the undergrowth is open. The shrubs are mostly soft mingimingi (*Cyathodes fasciculatus*).

11. (Next two pages) Hard beech forest on a level site, Tunnel Reserve, Te Marua, Upper Hutt. A large hard beech on the right with rounded buttresses. Note the deep leaf litter on the forest floor and the attractively layered foliage of the young beech trees.

Nevertheless the forests now are very similar in composition to those before the Ice Age, and share with them a number of species such as rimu, kahikatea, kauri, rewarewa and probably many others.

This is a great contrast with the situation at middle latitudes in the northern hemisphere where forests now are totally unlike those in much earlier geological times. For instance, in the strata under London, 80 million-year-old fossils were discovered of palms and other trees whose nearest living relatives now grow in Malaya. Severe Ice Age cold in the northern continents eliminated this type of forest, which was replaced by deciduous broadleaf forests at middle latitudes and cold-tolerant coniferous forests at higher latitudes.

Why did the pre-Ice Age forests persist, even though modified, in New Zealand? Probably because of New Zealand's narrow ocean-surrounded land mass and its remoteness from continents. Sea gains and loses heat more slowly than land, so that the strong oceanic influence over New Zealand would have reduced the severity of winters during cold Ice Age periods sufficiently for the ancient forests to survive in particularly favourable sites.

This effect of lessening of climatic extremes applies to the southern hemisphere generally as the oceans there greatly exceed the land areas. There are extensive southern beech forests in southern South America, including some deciduous species, and also in Tasmania and south-east Australia, as well as in the mountains of New Caledonia and New Guinea. Conifer broadleaf forests are to be found in central Chile, in south-east Australia and in a limited coastal area of south-east Africa. These southern land areas were once united in the continent of Gondwana which at this time had a mild to warm and moist climate and was largely forest covered. After the break-up and moving apart of the portions of Gondwana, forests were eventually eliminated from Antarctica by Ice Age cold and in varying degrees from Australia, southern Africa and southern South America, largely by the development of continental aridity.

In the chapters that follow the life styles of a range of forest plants—where and how they grow—will be explored. A sequence will be followed from the forest floor to the tree tops and special attention will be given to the many conspicuous plants with specialised lifestyles—the vines, epiphytes, parasites and saprophytes.

12. (Opposite) Mountain beech near Lake Ohau with flowers of the scarlet mistletoe (*Peraxilla tetrapetala*) providing a bright splash of colour.

13. (Above) The golf-ball-like fruiting bodies of the fungus *Cyttaria* parasitic on silver beech. Mt Robert, Nelson Lakes National Park.

21

2. The forest floor

The all important foundation of the forest floor is the soil. Soil begins to form with the weathering of parent material, which is most often solid rock, but sometimes other materials such as volcanic ash and sand. Most weathering is caused by rainwater which is weakly acidic from dissolved carbon dioxide from the air. If plants did not become involved, the rock fragments and minerals dissolved from them would simply be washed away and the rock surface steadily lowered. However, some of the minerals released by weathering are needed by plants, so they soon establish on the weathering rock surface. At first the soil layer is very thin and supports only very small plants such as lichens and mosses which hold the rock fragments and take up some of the dissolved minerals. Later, as a deeper soil develops larger plants appear, and eventually, climate permitting, forests establish.

Soil minerals taken up by plants are used to support growth. They are finally returned to the soil incorporated in leaves, twigs and other plant parts. The litter also contributes new organic compounds resulting from the breakdown of sugars, starch and proteins that green plants manufacture in their leaves.

Some plants require and absorb large amounts of minerals and as a consequence the litter they return to the soil is mineral-rich. This fertile litter is rapidly broken down into humus by decay organisms and the particles become incorporated in the upper layers of the soil. This is called 'mull' humus. The fertility of this type of soil encourages proliferation of soil animals such as worms, many insect larvae and mites that help break down the litter and also tend to spread minerals and humus throughout the soil. Worms are particularly good at this as they constantly bring soil up to the surface from below. These processes under nutrient-demanding trees result in a soil uniform in texture with little evidence of layering in profile. Most of the broadleaf species in the conifer broadleaf forests are mull formers.

Trees that require and therefore absorb few nutrients from the soil provide a litter low in nutrients that is not very 'palatable' to decay organisms and so breaks down only slowly. Such litter builds up to considerable thicknesses, and is known as a 'mor'. The ineffectiveness of the trees in absorbing mineral nutrients, and the increasing acidity of the slowly decomposing litter, leads to rapid leaching of minerals from the soil—including the iron and aluminium oxides that constitute clay. In extreme cases the soil layer below the litter and humus becomes quite pale from everything leaching away except relatively insoluble silica or sand. The iron and aluminium oxides are deposited at a lower, presumably less acidic, level in the soil to form a 'hard pan' which can become concrete-like and may impede drainage of the soil. Such strongly layered and infertile soils are called 'podsols'.

The conifers in the conifer broadleaf forest are mor formers. Slowly decaying litter often builds up around the bases of their trunks so they appear to stand on the summits of little hills. The southern beech species are even stronger mor formers and deep carpets of brown, yellow and red leaves are a conspicuous feature of their forests. But the kauri is the record breaker here. Kauri leaves, bark flakes, twigs and cone scales can build up to depths of more than a metre, especially near large trees, and the soils beneath are highly leached and infertile with impressive sandy leached layers above and hard pans below.

It has been suggested that despite the delaying action of even nutrient demanding plants the leaching process under forest climates eventually prevails and there is steady loss of soil fertility over a time span of several thousand years. On the very wet west coast of the South Island, vegetation sequences have been studied on dated terraces formed by the retreating of the Franz Josef Glacier from its maximum extent about 22,000 years ago. At the first stage a thin soil supports only small, herbaceous plants and shrubs. As a deeper, more fertile soil results from the breakdown of minerals of the parent material and the activities of certain microorganisms able to convert nitrogen into nitrates, eventually low forest develops, leading to dense, tall conifer broadleaf forest after about 5000 years. After about 12,000 years soil fertility gradually declines and by 20,000 years forest can no longer be supported, being replaced by a low cover of shrubs.

It should be noted, however, that in a mountainous geologically active country like New Zealand this sequence would often be interrupted and

14. Aerial view of the Olivine Range near Fiordland. The contact between ultramafic and schist rock is a diagonal line running from mid lower left to mid upper right. On the right there is the normal forest to alpine vegetation sequence with increase in altitude, to the left the infertility and toxicity of the ultramafic 'serpentine' soils reduces the vegetation at all altitudes to a sparse cover of shrubs and herbaceous plants. (Photo: P.F. Newsome)

taken back to the beginning by such events as land slides exposing new rock surfaces and by the burial of soils by alluvium and volcanic ash.

Soil parent materials vary in the amounts of minerals useful to plants that they contain; for example, among volcanic rocks, basalt is a much richer source of nutrients than rhyolite. The soils derived from most rocks are moderately to strongly acid, but those from limestone or marble, which are strongly alkaline, are usually also alkaline and this can influence the plant species present.

The extreme case of parent rock influence on soil and vegetation is that of ultrabasic or 'serpentine' rocks which are often very low in minerals useful to plants and high in certain metals such as nickel and chrome that are toxic to most plants. As a result localised occurrences of ultrabasic rocks in the north and south of the South Island and the far north of the North Island are occupied by stunted forest, shrubbery or open vegetation where tall forest can be found in the same regions on normal soils.[14] Some species (serpentine endemics) have evolved that are tolerant of the toxicity of serpentine soils and are restricted to them.

The Plants of the Forest Floor

The most abundant plants on the forest floor are ferns and this has led to New Zealand being sometimes referred to as 'The Land of Ferns'. Altogether we have about 200 species of ferns and related plants from tall tree forms up to 20 metres in height to tiny filmy ferns with translucent fronds only a centimetre or two long.

15. A dense ground cover of crown fern under beech and kamahi, Kaitoke Regional Park. The rather shrivelled-looking fertile fronds are at the centre of each crown.

16. (Opposite) Hen and chicken fern (*Asplenium bulbiferum*) showing the little plantlets that arise as outgrowths from cells of the upper frond surface. This is an alternative mode of reproduction as the fern also forms spores as is usual on the frond undersides.

24

17. (Opposite) King fern (*Marattia salicina*) with an uncoiling frond. King fern belongs to an ancient group of mostly tropical ferns. Our species grows in damp gullies in forest in the northern North Island, but is becoming less common as its large starchy rootstock is eaten by wild pigs.

18. (Above left) The kidney fern (*Trichomanes reniforme*) is one of our most unusual ferns most frequently seen in abundance on the floor and lower trunks of trees in beech forests. The forking veins can be seen clearly in the translucent fronds and the spindle-shaped clusters of sporangia are set around their rims.

19. (Above right) The crape fern or Prince of Wales feathers (*Leptopteris superba*) is much admired for its soft fluffy texture. Here it is growing near a stream in the Akatarawa Ranges near Wellington. Unfortunately it is disappointing in cultivation as it only grows at its best under cool, shady, constantly moist conditions.

20. (Right) The giant moss, *Dawsonia superba*, said to be the tallest moss in the world. Near Erua, central North Island.

Many ferns prefer moist shady sites so grow in gullies alongside streams or beside waterfalls, but others can grow on the open forest floor. Notable among the latter is the crown fern (*Blechnum discolor*)[15] a mini-tree fern that sometimes covers extensive areas to the exclusion of other plants. *Blechnum* is a world-wide genus with eighteen New Zealand representatives. The blechnums are unusual in having spore production restricted to special fronds that often have a dead and shrivelled appearance.

Another well represented genus is *Asplenium* and the best known of our native species is the 'hen and chicken fern' (*Asplenium bulbiferum*).[16] As well as reproducing by spores it also forms readily separating plantlets (the 'chickens') on the fronds' upper sides.

Filmy ferns also are strongly represented and on the forest floor are probably most common on rotting logs. The fronds of these ferns are only one cell thick between the veins and so appear very delicate and translucent. They are not so delicate as they look, however, as rain quickly revives them from their shrivelled state induced by occasional droughts in summer. Most filmy ferns have deeply divided fronds. The notable exception is the kidney fern[18] which has relatively large kidney-shaped fronds that are both striking and attractive when growing en masse.

The two species of *Leptopteris* also have very thin fronds and look like large versions of the unrelated filmy ferns. The Prince of Wales feathers or crape fern (*Leptopteris superba*)[19] is a plant of cool, moist shady places and has very attractive fluffy textured fronds.

One of the attractively fan-like umbrella ferns (*Sticherus cunninghamii*)[21] may also form colonies in places.

Mosses and Liverworts

The mostly much smaller mosses and liverworts are even more abundant in New Zealand forests, with about 1000 species, and are strongly dependent on shady, constantly moist sites for their survival. Only a few mosses, including the giant moss (*Dawsonia superba*)[20] up to 60 cm high, can grow in more open sites on the floor of drier forests.

22. A leafy liverwort (*Schistochila*) with its distinctive sporangia: translucent whitish stalks and black heads that contain the spores.

Unlike the ferns with their large compound leaves, mosses and liverworts have very small, simple and delicate leaves. Some liverworts in fact do not have leaves at all, but consist of flat sheets of cells that gradually spread over the substrate. An especially large New Zealand example is *Monoclea* which forms extensive dark-green masses in very moist and shady places alongside streams or waterfalls.

The leafy liverworts[22,37] are not easy to distinguish from mosses except when they are bearing their spore-producing organs or sporangia. The sporangia look rather like matches with delicate translucent stalks and black heads. The heads are termed capsules and when ripe they split into four valves to release the spores. In mosses the sporangia have slender wiry stalks and the capsules are green and initially are partly enclosed by a loose pointed cap. The moss capsule opens by a round lid at the tip and the spores sift out through the resulting hole.

Fungi

Conspicuous with their often bright and strange colouration among leaf litter are the fungi, especially in the moister seasons of autumn and winter. Nowadays fungi are often treated as a kingdom separate from plants and animals. They have firm cell walls like plants, but, like animals, they have no chlorophyll and so are unable to manufacture their own food materials. As a consequence both fungi and animals have to obtain their food either directly from green plants or indirectly via animals that are part of a food chain leading back to green plants.

Most of the time the fungi are not visible and consist of pale branching threads of cells that absorb nutrients from living green plant or sometimes animal tissue or from decaying plant litter on the forest floor. In the former case the fungi are called parasites and in the latter saprophytes.

It is only when reproduction is initiated that the solid, although soft, and visible 'fruiting bodies' are formed in some of the fungi. We probably have several thousand species of these larger fungi in New Zealand. Because of the edible mushroom the most familiar are the gill fungi. Here the fruiting body consists of a stalk and an umbrella-like head beneath which are the delicate gills that form the spores.[23] Among our many native examples growing among leaf litter or on rotting logs there are a number that catch the eye with their bright colours.

23. (Opposite) A bright red gill fungus (*Hygrocybe rubro-carnosa*) growing in leaf litter.

24. (Below) A brown, almost scaly forest puffball (*Lycoperdon pyriforme*).

25. (Left) A striking purple pouch fungus (*Thaxterogaster porphyreum*) among beech litter.

26. (Opposite) Bright orange discs of *Peziza* growing on a mossy bank.

27. (Above left) A mummified cicada larva with spore-producing fungal tissue growing out from its head.

28. (Above right) Stinkhorn fungus *(Aseroe rubra)* with flies attracted by the smelly fluid containing spores.

29. (Right) Another group of the stinkhorn fungus with some of the soft, egg-like structures from which they emerge.

Other fungi of mushroom and other shapes form their spores in close-set vertical pores—the polypores. In yet another group, which includes the puffballs,[24] the spores are contained in a bag-like structure, with[25] or without a stalk. With the puffballs the spores puff out through a hole at the top even with the slight pressure of wind gusts or rain drops. In other cases the spores are only released when their container rots away. A remarkable member of this group is the 'stinkhorn'.[28,29] The fruiting body opens into a flower-like structure containing a spore-filled evil-smelling fluid.

Another common group of fungi have the fruiting body in the form of a sometimes brightly coloured disc or cup.[26]

Very curious fungi, which in fact straddle the boundary between fungi and animals, are the slime moulds.[30] Here the plant body is an amoeba-like mass of protoplasm without cell walls which slowly oozes over suitable surfaces. Only when spore production takes place are cell walls formed.

Two fungus-caused curiosities in New Zealand are the vegetable caterpillar and the vegetable cicada.[27,31] Specific fungi attack and mummify underground moth and cicada larvae respectively and eventually fungal stalks grow out from the heads of the larvae above ground level where spores are formed.

Fungi also form a very important association with the roots of probably the majority of seed plants in forests. The fungal threads penetrate the cells of root tips and cause them to assume a rather short and stunted shape. Such roots are called mycorrhizae ('fungus roots'). Both fungus

30. A slime mould that has spread over a piece of wood and adjacent leaf skeletons.

and seed plant benefit from this association. The fungus gains access to carbohydrates—sugar and starch—and the roots obtain water and mineral nutrients from the far extending fungal threads.

Lichens

What one can see of lichens externally is fungus, mostly belonging to the group of fungi with disc-like fruiting bodies. But the fungal components of lichens have solved the problem of not being able to manufacture carbohydrates by incorporating within their tissues a layer of unicellular green plants belonging to the group of aquatic plants known as algae. This is a mutually beneficial situation: the algal component is provided with a sheltered moist environment and the fungal component shares the products of the alga's chlorophyll. Lichens are not regarded as very highly evolved, but they are a versatile and hardy group being represented in just about every terrestrial habitat. On drier forest floors lichens are not very common, but in wet forests at higher altitudes[32] or in the west they are often abundant particularly on rocky outcrops, logs and exposed roots of trees.

There are three main forms of lichen.[33] Crustose lichens form very thin layers on smooth rock or bark and become more or less incorporated into the substrate. They are usually circular and grow very slowly at their edges. Foliose lichens are loosely attached sheets of cells not unlike the non-leafy liverworts in form. Fruticose lichens are erect and branching like mini-shrubs. It is estimated that there are about a thousand native lichen species in New Zealand, although only some of these are forest inhabitants.

31. (Above) Splatters of white spores from the emergent tips of the fungus *Cordyceps sinclairii* parasitic on buried cicada larvae.

32. (Next two pages) Interior of higher altitude mountain beech forest, with abundant lichens on rocky outcrops and trunks. Upper Travers Valley, Nelson.

33. (Page 38) Crustose, foliose and fruticose lichens on a branch. Huxley Valley cloud forest. Southern Alps.

34. (Previous page) *Astelia chathamica*, the sole species of the genus on the Chathams where it grows on the forest floor. Owing to its silvery leaves and bright orange berries this is becoming popular as a garden plant. Otari Native Botanic Garden, Wellington.

35. (Above) The remarkably attractive leaf mosaic of parataniwha (*Elatostema rugosum*), Mangamuka River near Kaitaia. There are many species of *Elatostema* in the Asian tropics. Our species is most common in the north of the North Island, but there are localised occurrences further south in the island.

36. (Opposite) Brilliant blue-purple berries and flowers of *Pratia physaloides*. Mostly restricted to shady streamsides in the far north of the North Island. Otari Native Botanic Garden, Wellington.

Flowering Plants

Although they are not as conspicuous as the ferns, a range of herbaceous flowering plants are to be found on the forest floor. Some are colourful in neither flower nor fruit; these include the greyish green bush rice grass and species of the sedge genus *Uncinia* whose hooked seeds readily attach to clothes and somewhat painfully to hairy legs.

Other herbaceous flowering plants, however, are colourful in foliage, flower or fruit. Large tussocks, often with colourful berries, are provided by several species of *Astelia*[34] belonging to the lily family, the most notable being the 'kauri grass',[68] often dominant in the ground cover of kauri forests.

In the North Island the parataniwha (*Elatostema*)[35] can form curtain-like mosaics of foliage on shady, moist banks. The flowers are minute, but the reddish to irridescent hues of the leaves justify the early common name—New Zealand begonia.

An even more striking plant is a lobelia relative—*Pratia physaloides*.[36] It is restricted to the far north of the North Island and there often to deep shaded gullies alongside streams. The ascending stems have quite large soft leaves and at first sight there seems to be nothing remarkable. Separating the leaves, however, may reveal long flowers of lobelia form and hanging berries 1-2 cm in diameter, both coloured a brilliant blue-purple.

More modest in size but more widespread are the nerteras,[37] herbaceous relatives of the coprosmas. Their slender stems and small leaves

form close mats decorated in season with red or orange berries like shiny beads.

Outstanding for their elaborate flowers are the ground orchids. Most die down in winter to bulbs but some die down in summer instead. The spider orchids (*Corybas*) generally have a single broad leaf and a relatively large purplish flower with long 'antennae' from the petals. The greenhood orchids (*Pterostylis*)[38] are taller and more leafy and the flowers may be quite a bright green with pink tipped 'antennae'. Some of the ground orchids completely lack chlorophyll and so live as saprophytes. All of the gastrodias[39,40] lack chlorophyll and are quite strange in appearance. Their slender stems, which can be up to a metre high, have a polished wood-grain appearance, and their knobbly flowers seem only half open. In colour these plants range from cream to almost black. *Thismia* is an orchid relative also lacking chlorophyll, and its small reddish flowers among leaf litter have been described as 'small, red lanterns'.

Shrubs

A number of small shrubs also inhabit the forest floor. Some are very small-leaved and twiggy, such as *Coprosma rhamnoides* and some of its relatives, and the soft and prickly mingi mingis (*Cyathodes*) of some beech forests. Others have much larger leaves and branch less freely. The genus *Alseuosmia* ('perfume of the grove') is prominent here. The name refers to the strong pleasant perfume of their flowers in spring and the several species are best represented in the northern North Island. *Alseuosmia pusilla*[41] extends to the northern South Island and is a very small often unbranched shrub. It bears eye-catching clusters of berries that are like bright red wax.

A distinctive small shrub in northern parts is *Rhabdothamnus solandri*.[42] Its leaves have a sandpapery feel to them and the flowers are tubular and two-lipped and coloured from orange to brick-red.

This is probably the appropriate place to mention shrubs that grow on banks and cliffs in the flood zone of streams. Perhaps their site could be considered as vertical forest floor! In the tropics there are communities of plants that grow in this zone and they tend to have tough stems and narrow streamlined leaves to resist water battering during floods. Plants of this habitat in New Zealand have not been closely studied from this point of view, but there are clearly several shrubs suited to these special conditions.

Several leafless brooms (*Carmichaelia*)[43] grow in the flood zone and their tough, narrow, flattened mostly leafless stems would seem well suited to survive flooding. Several narrow-leaved hebes occur here too, as well as the appropriately named *Parahebe catarractae*.

37. (Opposite) *Nertera ciliata* with whiskery berries growing amongst the leafy liverwort *Plagiochila* on a moist bank.

38. (Above) *Pterostylis banksii*, the largest of our greenhood orchids. The petals are drawn out into pink-tipped 'antennae'.

39. (Opposite) A tall flowering spike of the saprophytic orchid *Gastrodia cunninghamii* in beech forest on the Five Mile Track, Orongorongo Valley, Wellington. The plants all lack chlorophyll and range in colour from almost black, as here, to yellow. The underground parts are associated with fungal threads that are linked with tree roots so *Gastrodia* may be indirectly parasitic.

40. (Above) Close view of the strange flowers of an attractive yellow form of *Gastrodia cunninghamii*. Onamalutu Reserve, Wairau Valley.

41. (Above left) The diminutive forest floor shrub *Alseuosmia pusilla* in the Akatarawa Ranges near Wellington. The leaves mimic those of the mountain horopito, but the waxy red berries are quite distinctive.

42. (Above right) Tubular, orange-red flowers of the small shrub *Rhabdothamnus solandri*.

43. (Above) The leafless broom *Carmichaelia odorata* in the flood zone of a vertical river cliff, Kaitoke Regional Park, Upper Hutt. Accompanying it are a narrow-leaved koromiko (*Hebe stricta*) and a variety of small ferns and mosses.

Special features of tree roots and trunk bases

In the tropics a number of trees, mostly growing in swampy sites, have extensive, very thin, supporting buttresses at their trunk bases. Because they are so thin they are called plank buttresses. Most of our New Zealand trees are without buttresses or have small, rounded ones as in mature trees of kahikatea and tawa. Pukatea, however, provides a good example of plank buttresses.[44,45] It is a tree of swamp forests or grows close to streams in other sites. It can have up to five or six buttresses that are triangular in form and may extend out from the trunk proper for several metres.

Although usually not favouring swampy sites, large trees of red beech sometimes have large buttresses[46] that may also extend for several metres, but they seem not to be so planklike as some formed by pukatea.

The other problem faced by swamp trees is the lack of oxygen for the living tissues of the root systems. This is often overcome in warmer latitudes by the growth of special roots or pneumatophores ('breathing roots') above the surface of the swamp. The pneumatophores generally have pustules of loose corky cells through which air can be taken in.

In New Zealand we have several examples of trees with pneumatophores. In pukatea a young branch root that is going to become a pneumatophore grows above the surface of the swamp then loops back in. The loop gradually becomes woody, extends vertically upwards and eventually assumes a shield-like form.[47] Pukatea pneumatophores can sometimes be a metre high.[48]

44. (Below left) A strikingly plank-buttressed pukatea (*Laurelia novae-zelandiae*) near a stream in Otari Native Botanic Garden, Wellington.

45. (Below right) A particularly thin and sharp-edged buttress of pukatea, Hemi Matenga Reserve, Waikanae.

46. (Above) A large red beech (*Nothofagus fusca*) with impressive buttresses. Kaitoke Regional Park, Upper Hutt.

47. (Left) A shield-like pneumatophore (breathing root) of pukatea in swamp forest at Nga Manu Sanctuary, Waikanae. The pustules of loose corky cells where air enters can be seen on its surface.

48. (Opposite) A group of pukatea pneumatophores, some of which are a metre high. These are derived from a series of loops on the same root extending many metres from the trunk in the left background. Nga Manu Sanctuary, Waikanae.

49. (Page 50) Branching finger-like pneumatophores of swamp maire (*Syzygium maire*). Nga Manu.

50. (Page 51) Finger-like pneumatophores of the mangrove (*Avicennia resinifera*) projecting above the mud at low tide. Paihia, Bay of Islands.

Another but smaller swamp forest tree—the swamp maire (*Syzygium*)—forms pneumatophores of a different type. Here the young root that grows above the surface does not loop back, but remains vertical and finger-like and eventually branches to form a coral-like cluster.[49]

Our sole mangrove, growing in muddy estuaries and harbours along northern North Island coasts, also has finger-like pneumatophores but here they do not branch. The thousands of close-set pneumatophores of the mangrove exposed at low tide make a strange and impressive sight.[50,51]

Very curious root formations of several New Zealand trees are known as 'wooden roses'. These are caused by a flowering plant parasite, the pua reinga (*Dactylanthus taylori*). The parasite has no chlorophyll and most of the time it is hidden under leaf litter. It becomes visible only when the well camouflaged purple-brown, scaly flower heads emerge.[52] The body of the parasite is ball-like and is attached at the end of a tree root that expands into a disc-like form. The nutrient-absorbing structures of the parasite make radiating grooves into the surface of the disc and when the parasite is removed the 'wooden rose'[53] is revealed.

51. (Page 52) A tidal channel in mangrove forest, Paihia, Bay of Islands. The pneumatophores are most abundant at the edge of the channel.

52. (Page 53) The strange scaly inflorescences of the root parasite pua reinga (*Dactylanthus taylori*).

53. (Left) The 'wooden rose' formation of the host root revealed when the *Dactylanthus* parasite is separated from it after boiling.

3. Shrubs, small trees and tree trunks

Shrubs and small trees

With the exception of a few small shrubs, the plant cover of the forest floor is dominated by soft herbaceous plants. Moving upwards, however, woody plants reign, from slender-twigged shrubs to massively woody giants such as the kauri. In this chapter we will consider the zone from about two metres to 12 metres above the forest floor, which includes shrubs and small trees below the crowns of taller trees, as well as the trunks of the latter passing through.

In common with most New Zealand plants the flowers of the majority of these shrubs and small trees are neither large nor colourful. Some are pure white and so striking en masse, others are small and greenish yellow and pass unnoticed. Butterflies and long-tongued bees, as pollinating agents, are attracted to brightly coloured and perfumed flowers and it is suggested that their original paucity in New Zealand is the reason for the modesty of so many of our flowers. Perhaps in compensation, many of them are succeeded by brightly coloured berries whose contained seeds are distributed in the droppings of birds that feast on them.

The relatively large bright red or yellow flowers of the kowhai, some rata, some mistletoes and others are bird-pollinated as birds are attracted by such colours.

54. Kawakawa (*Macropiper excelsum*) with its distinctive heart-shaped leaves and candle-like clusters of orange berries.

In conifer broadleaf forest of milder climates there are several shade-tolerant species which seem to be at their best under a completely closed canopy. Kawakawa has attractive heart-shaped leaves, prominently jointed stems and, in season, small berries clustered together to form small orange 'candles'.[54] Kawakawa is related to the pepper plant of the tropics and is sometimes known as 'pepper tree' because of its hot tasting leaves. It is also related to the similarly named kava plant of Fiji.

Horopito is another shade and moisture loving species. It has blackish twigs and very shiny, dark-green leaves which may be whitish below. The flowers are small, but the berries are bright orange.

Hangehange may also be common. It can be mistaken for a coprosma with its simple opposite leaves but, among other differences, the fruits are dry, two-valved capsules, not attractively coloured berries.

Other shrubs and small trees are found in subcanopy shade, but may be equally or more abundant in moist, sheltered, but better-lit canopy gaps caused by the collapse of large trees. They may also play a role in forest regeneration in larger open areas resulting from such events as landslides and gales. Prominent here as small trees are the mahoe with its brilliant purple berries,[55] pigeonwood with bright orange berries,[56] wineberry with its finely toothed leaves and small but attractive red flowers[57] and putaputaweta with clusters of pure white flowers. Also frequent are our several kinds of tree fern,[58,59] sometimes forming dense groves, as well as our sole palm, the nikau,[60,61,62,63] where the climate is particularly mild. In moist sites, often near streams, the tree fuchsia[64] catches the eye with its orange-brown bark separating in papery strips.

55. (Opposite) The striking purple berries of mahoe (*Melicytus ramiflorus*) attached to woody twigs. The berries are eaten by parakeets and other birds.

56. (Below left) Bright orange berries of pigeonwood (*Hedycarya arborea*). The circle of berries derived from each flower appear to be attached to spokes of a wheel.

57. (Below right) Wineberry or makomako (*Aristotelia serrata*) has masses of small but attractively coloured flowers in the spring.

58. (Previous two pages) The mamaku or black tree fern forming a dense grove on a moist, shady hillside. The mamaku is our largest tree fern—up to 20 m. It is notable for the jet black stalks of its leaves. Near Wainui Waterworks Reserve, Wellington.

59. (Opposite above) Uncoiling fronds of the ponga tree fern (*Cyathea dealbata*).

60. (Opposite below) Handsome nikau palms (*Rhopalostylis sapida*) at the edge of a coastal forest behind a beach at the mouth of Big River, north-west South Island. Note on the right, a profuse growth of the long-leaved climber kiekie (*Freycinetia banksii*) which has reached the fronds of one palm. (Photo: Craig Potton)

61. (Above) Interior of a grove of young nikau palms, Nikau Reserve, Paraparaumu.

62. (Next page) Nikau palm with a pink-flowered inflorescence above and a mass of red fruits below which slowly developed from the flowers of the previous year. Note the pronounced scars on the trunk where the encircling leaf bases were originally attached. Wellington Botanic Garden.

63. (Following page) Close view of a nikau inflorescence with the spathe that originally enclosed it persisting above. Westhaven Inlet, north-west South Island.

61

64. (Opposite) Flowers of the tree fuchsia or kotukutuku (*Fuchsia excorticata*) growing out from the trunk. The attractive orange-brown bark separates in tissue-thin strips. Flowers awaiting pollination by birds such as the tui are purple and green with strikingly blue pollen. After pollination the flowers turn red.

65. (Above left) Furry capsules of the titoki (*Alectryon excelsus*). Before they open the capsules have a distinctive helmet shape. They open after the fashion of a bi-valve shellfish to reveal a single, large, shiny black seed partly enclosed by a convoluted red aril that provides food for native pigeons and other birds.

66. (Above right) Cabbage trees (*Cordyline australis*) and *Phormium tenax* in a swampy site. Doubtless Bay, northern North Island. (Photo: John Dawson)

67. (Left) Golden tubular flowers of the common kowhai (*Sophora microphylla*). The tubular form of the flowers allows storage of sufficient nectar to attract tuis and other bird pollinators. The seed pods of the kowhais are unusual with their four wings and constrictions into one-seeded segments.

Somewhat smaller trees are pate with its large compound leaves divided into a fan-like arrangement of leaflets and the large-leaved coprosma, kanono.

Other species are largely restricted to the more open sites—several pittosporums, including tarata and kohuhu, kaikomako, the bubbly-leaved ramarama, two large-leaved coprosmas (karamu and shiny karamu), five-finger, lancewood with its striking contrast between juvenile and adult, titoki with its unusual and colourful fruit,[65] two 'daisy trees' (the large-leaved rangiora and heketara), and the lowland[66] and mountain cabbage trees.

The two tree kowhais are often near streams or lake edges. They are notable for their large, tubular bright yellow flowers.[67]

Other forest types have at least some different subcanopy small trees and shrubs, while low altitude beech forests in milder regions are often quite open below the canopy with few or no tall shrubs or small trees. In kauri-dominated forests[68] mairehau, the narrow-leaved neinei and the pure white flowered daisy shrub *Brachyglottis kirkii* are often conspicuous.

Other subcanopy shrubs or small trees appear in both conifer broadleaf and beech forests at higher, cooler and wetter altitudes. They include: the mountain horopito[69] with its attractive red and yellow blotched leaves that seem to be in a permanent state of autumn colouration; stinkwood—a species of coprosma (*C. foetidissima*) whose leaves have an unpleasant smell when crushed (the name given to this species by Cook's botanists could be translated as 'stinking dung plant'!); mountain five-finger and its smaller-leaved relative haumakaroa; and the small tree broadleaf.

Tree trunks

Tree trunks have an outer layer of corky bark that insulates the delicate conducting and dividing cells, near the junction of wood and bark, from cold and drought. In moist tropical forests near the equator insulation against cold and drought is unnecessary so the majority of tropical trees have thin, smooth bark. By contrast trees of temperate regions with severe winters generally have thick and rough bark.

The New Zealand forests lie between these two extremes. A quick assessment suggests that about 63 per cent of New Zealand trees have fairly smooth bark including species of genera widespread in the tropics, for example, kohekohe, tawa and titoki; 19 per cent have rough, furrowed bark including the beech[70] species, the maires, pohutukawa, ngaio and the cabbage tree; 10 per cent have bark separating in hard flakes, large flakes in the case of rimu[71] and small, round flakes in kauri, matai[72] and others; and 8 per cent have the bark separating in long, thin strips including kanuka, Hall's totara, the cedars,[73] neinei, the tree fuchsia and several daisy trees (*Olearia*). With the rough-barked trees often the bark of young plants is smooth.

68. (Opposite) Scene within a kauri-dominated forest at Puketi near the Bay of Islands. The undergrowth includes kauri grass (*Astelia trinervia*), neinei (*Dracophyllum latifolium*) and *Brachyglottis kirkii* var. *augustior*. (Photo: John Dawson)

69. (Next page) Mountain horopito (*Pseudowintera colorata*) showing its permanent 'autumn colours'. Mt Taranaki.

70. (Following page) Rough, furrowed bark of adult hard beech (*Nothofagus truncata*).

The rough-barked trunks and major branches of pohutukawa are notable for producing an abundance of freely branching aerial roots.[74] Some grow down the trunks to the ground, others hang down from the branches and divide freely at their lower ends to assume a straw broom-like appearance. This is a very similar habit to that of some of the banyan figs in the tropics. In the preferred habitat of the pohutukawa on far northern coastal cliffs the aerial roots must be useful in providing extra anchorage. The kamahi too, in high rainfall areas, often forms aerial roots on its usually several trunks. Here also the tips of these roots often branch into a mass of fine rootlets that have an almost fur-like texture. In some cases they resemble the paws of a large animal.[75] When a kamahi is blown over these roots anchor into the ground while the upper sides of the trunks form a mini-forest of new branches.[76]

Production of flowers and later fruits directly on trunks and major branches (cauliflory) is a feature of some tropical trees. In some cases the flowers are pollinated and the seeds are dispersed by bats which can echolocate flowers and fruits more readily in the open spaces below the forest canopy, but birds are also involved.

In New Zealand we have several examples of this, the most notable being the kohekohe. Kohekohe has many tropical relatives that also exhibit cauliflory. Our species can be abundant in forests near the sea through the North Island to the northern tip of the South Island. In early winter the trunks and branches are festooned with sprays of waxy white flowers quite like orange blossom in appearance.[77,78] At about the same time the large green fruit capsules from the previous year are

71. (Page 70) A rimu trunk (*Dacrydium cupressinum*). The large bark flakes leave behind an attractive 'water-mark' pattern.

72. (Page 71) The small, round bark flakes of matai (*Prumnopitys taxifolia*) leave a hammer-mark pattern on the trunk. The scars of recently fallen bark flakes can be quite a bright red as here.

73. (Page 72) The trunk of a kawaka (*Libocedrus plumosa*) looking like a sheet metal sculpture. Radar Bush near Cape Reinga.

74. (Page 73) 'Straw broom' aerial roots on a pohutukawa (*Metrosideros excelsa*). The new root growth is often red coloured.

75. (Opposite) Furry, paw-like aerial roots of a fallen kamahi (*Weinmannia racemosa*). Waikanae.

76. (Left) A mini-forest of upright branches from a fallen kamahi. Akatarawa Ranges, Wellington.

splitting open to reveal three large seeds each enclosed by edible fleshy orange arils.[79] The tui has been observed taking nectar from kohekohe flowers and also feasting on the arils.

As well as woody trunks there are also a variety of tree fern trunks. These differ in several respects from the trunks of ordinary trees. The latter have a special layer of dividing cells inside the bark which adds a new layer of wood to the trunk each year thus enabling it to support the steadily enlarging crown. Tree ferns do not have such a wood-forming layer, but they have other ways of enlarging their trunks to support their often very large and heavy crowns of leaves[80]—the stumps of the old leaves persist and add considerably to the thickness of the

stem, and, when the trunk has attained some height, its base, which takes much of the strain, is greatly enlarged by the outgrowth of a fibrous entanglement of wiry roots.[81] The most impressive example of trunk forming in this way is provided by the wheki-ponga. Here the whole trunk becomes enclosed in a deep layer of very fine fibrous roots which greatly exceeds the diameter of the stem proper deep within.

77. (Opposite) Inflorescences festooning a major branch of a kohekohe (*Dysoxylum spectabile*).

78.(Right) Close up of the waxy, orange-blossom-like flowers of kohekohe.

79. (Below) The large capsules of the kohekohe split open to reveal three large seeds partly enclosed by smooth, orange, fleshy arils.

80. (Page 78) Twin ponga tree fern trunks (*Cyathea dealbata*) with persistent leaf bases above and an outgrowth of wiry roots below binding the bases of the two trunks together.

81. (Page 79) Ponga tree fern trunk cut through near the base. This dramatically illustrates that the base of a tree fern trunk is largely composed of interlaced wiry roots. The cavity at the centre is where the true stem tapered to a point at ground level.

4. The forest roof

The forest roof or canopy is formed by the crowns of taller trees. In New Zealand, where conditions for tree growth are at their best, the canopy is 20-25 m high, although forests dominated by hard, red and silver beech may attain 30 m. When conditions are less than ideal the canopy is lower, as a result of either decreasing temperature with increase of altitude or latitude, or increasing wind and salt exposure near the sea.

Emergents

In many tropical forests there are also scattered, exceptionally tall trees standing above the canopy. These are known as emergents. Such trees establish early in a forest's history when there is still adequate light at ground level and they may for a time form a continuous layer. As more shade-tolerant trees grow up below the emergents, the forest floor becomes too shaded for them to regenerate and as old trees die they are replaced only sporadically in the better light of larger canopy gaps.

In New Zealand the conifer broadleaf forest often has emergent trees, but most beech forests do not.

In the far north the kauri is an emergent of monumental proportions.[82] When young, it has an attractive narrowly conical form, but when mature it has a rounded 'stag-headed' crown supported by an immense cylindrical trunk up to 7 m in diameter. Exceptionally the kauri can attain a height of 60 m.

Other emergent conifers grow throughout the country and mostly belong to the southern family Podocarpaceae. Rimu[83] is the most abundant of these, mostly in better drained forests, but also on wetter sites in Westland. The juvenile stage of the rimu is particularly attractive with its graceful weeping habit,[105] but, although much less marked, even the adult branchlets hang downwards. The leaves are small and scale-like and en masse they appear yellow-green. The black seeds are borne singly and are seated on a red, fleshy berry formed by an aggregation of modified scale leaves. Tuis and other birds are attracted by the red berries and swallow the seeds as well. The latter mostly pass through the birds unharmed and are so dispersed over quite a wide area. The rimu can attain a height of 35 m or sometimes more. It is most common in conifer broadleaf forest, but may also be scattered through some lower altitude beech forests.

The kahikatea[139] rivals the kauri for being New Zealand's tallest, if not most massive, tree. Heights up to 40 m are frequent, but some older specimens in favourable sites can grow much higher. Kahikatea prefers moist sites, near streams in better drained forests, and it often dominates swamp forests. Like the rimu it has scale leaves, but its foliage tends to be grey-green with the branch tips directed upwards. The seeds and associated berry-like bases[84] may be abundant in good years, imparting a red to orange glow to the tree crowns.

82. (Opposite) A group of enormous kauri trunks (*Agathis australis*) in the Cathedral Grove, Waipoua Forest.

81

Totara is not so tall as kahikatea, but generally has a more massive trunk. Its leaves are needle-like, rather than scales, but like kahikatea it prefers alluvial sites, although well drained ones. Indeed totara is fairly drought-tolerant. Hall's totara and mountain cedar attain only about 20 m in height, but where they occur in higher altitude forests they are emergent above a low canopy. They are most common in montane conifer broadleaf forests, but mountain cedar, with its attractive conical form, may also be scattered through some higher altitude beech forests. Hall's totara has the seed plus berry pattern of other podocarps, but mountain cedar belongs to the cypress family and has small, dry and scaly seed cones.

Some broadleaf or flowering trees are also emergents. Rewarewa is often prominent in younger forests on hill slopes with its tall Lombardy poplar form reaching heights up to 30 m. Its leaves are stiff and coarsely toothed and its spikes of crimson flowers in spring are notable for their spirally coiled petals.[85]

Pukatea often accompanies kahikatea in moist places and may be co-dominant in swamp forests. Northern rata, in better drained forests, attains its height of 30 m or more by beginning life as an epiphyte or perching plant in the branches of another emergent tree, often a rimu. Its 'trunk' is made up of coalesced roots. Northern rata[86] shares a number of features with its relatives southern rata[87] and pohutukawa—broadly spreading crowns and red, bird-pollinated flowers with prominent stamens.

All the broadleaf emergents reach the South Island, but do not extend so far south as the conifers.

Canopy trees

Several species of conifer also contribute to the closed canopy of the conifer broadleaf forest, although they are generally less common than broadleaf trees.

The related miro and matai are found throughout, including Stewart Island. Both have short, narrow leaves, but the foliage of miro is the more attractive with its pointed leaves flattened in two rows to give a fern-like appearance. Both have 'hammer-marked' bark and seeds with fleshy coats. The seeds of matai are black; those of miro are larger and red.[88]

Three other canopy conifers—monoao, kawaka and tanekaha—grow in the northern, but not southern, North Island, although the last two reappear in the northern South Island. Monoao is a podocarp and shares with related smaller species a juvenile form with spreading needle leaves that changes without transition to a scale-leaved adult state. Kawaka with its flattened scale-leaved foliage looks more like a northern than southern hemisphere conifer and, indeed, along with its

83. (Opposite) A tall rimu (*Dacrydium cupressinum*) on a river terrace, Kaitoke Regional Park, Upper Hutt.

84. (Above) The seeds and berry-like stalks of kahikatea (*Dacrycarpus dacrydioides*).

85. (Next page) A close view of the unusual rewarewa flowers (*Knightia excelsa*). The four petals in each flower coil up into tight spirals reminiscent of Maori art.

more montane relative kaikawaka (mountain cedar), it belongs to the largely northern hemisphere cypress family. Tanekaha, although a podocarp, belongs to a group of unusual conifers ('celery pines') that appear to have somewhat fan-shaped leaves. The latter are not truly leaves at all, but short, flattened branchlets that develop in the angles between the true but very small scale leaves and the stems.

Although there are a number of broadleaved canopy trees, a few of them tend to dominate certain forests. Of the related tawa and taraire, taraire dominates many forests in the north of the North Island. Tawa may also be present as a minor, although often taller, component. When taraire drops out at about 38°S, tawa takes over as the dominant in many forests as far as the northeast of the South Island. Both these trees produce dark purple fruits like small plums, but are very different in their form of leaf—taraire leaves are broad and thick, tawa leaves are narrow, willow-like and relatively thin.

South from about 38°S in the North Island kamahi generally replaces tawa at higher altitudes and at lower altitudes as well in the north, west and south of the South Island and Stewart Island. In conifer broadleaf forests it often dominates the canopy, but where it occurs in wetter beech forests it contributes to the subcanopy. The often multi-trunked kamahi bears coarsely toothed leaves and the small cream-coloured flowers are arranged on distinctive and abundant candle-like spikes[89] standing above the foliage. The small capsules that follow the flowers are often reddish-brown.[90] Towai, a tree closely related to kamahi, is found north of the latter's range on the Coromandel and Northland peninsulas.

86. (Previous page) A northern rata (*Metrosideros robusta*) 'on fire' with bright red flowers. Te Marua lakes, Upper Hutt.

87. (Above) Brilliant flowers of southern rata (*Metrosideros umbellata*), Auckland Islands. (Photo: Barry Snedden)

Other canopy broadleaf species can be dominant near the sea.
Pohutukawa, with its broadly spreading crowns and bright red flowers
in summer, is often abundant along northern coasts of the North
Island, while karaka[91] with its thick, dark green leaves and ngaio, whose
leaves are spotted with oil glands, range much further south. Kohekohe
with large compound leaves can be abundant near the sea as far south
as the northern South Island. Sometimes it shares dominance with
tawa and in this case the canopy tends to be uneven with tawa crowns
occupying high points and kohekohe filling the spaces between.

Mangroves with their tolerance of salinity are a special case. When
conditions are favourable they can form a canopy of small trees[92] in the
shallow sea of northern North Island estuaries and harbours.

Other broadleaf species, although frequent and often widespread, do
not dominate the canopy. Hinau has somewhat rolled adult leaves with
a satiny lower surface and prominent pouch-like cavities (domatia)
alongside the midrib. Its flowers,[93] and those of its smaller relative
pokaka,[94] are distinctive in having white petals with deep incisions, as if
they have been cut by scissors. Black maire has narrow dark green leaves
and red berries. It and its smaller relatives are related to the olive tree of
the Mediterranean. Southern rata, although having localised occurrences in the North Island, is most abundant in the west and south of
the South Island and Stewart Island. In a good flowering season it is
even more spectacular than its relatives, northern rata and pohutukawa,
its leaves being completely hidden by bright red flowers.[87] Puriri is a
spreading tree of the north of the North Island in coastal and inland
forests. It bears tubular red flowers[95] and red fruits most of the year.

88. Attractive miro
(*Prumnopitys
ferrugineus*) foliage with
large seeds each
enclosed by a fleshy
coat.

89. (Above left) Flower spikes of kamahi (*Weinmannia racemosa*).

90. (Above right) Spikes of red-brown young capsules of kamahi (*Weinmannia racemosa*).

91. (Right) A karaka (*Corynocarpus laevigatus*) with large, fleshy fruits.

92. (Opposite) Scene near a tidal channel in a low mangrove forest, Paihia, Bay of Islands. The exposed mud at low tide is covered with finger-like breathing roots and mangrove seedlings.

93. Hinau flowers (*Elaeocarpus dentatus*) with their distinctively incised petals.

94. Typical adult foliage and flowers of pokaka (*Elaeocarpus hookerianus*).

95. Puriri (*Vitex lucens*) is a northern North Island tree with handsome palmately compound leaves and furry tubular red flowers. Puriri has many tropical relatives.

96. Black beech (*Nothofagus solandri*). Male flowers with red anthers and scaly leaf buds above.

97. Inflorescence buds of tarata or lemonwood (*Pittosporum eugenioides*) with many closely overlapping bud scales.

Finally, the beech forest canopy is often dominated by just one species of beech or sometimes two together. The lowland beeches—black and hard beeches—grow in the altitudinal zone favoured by conifer broadleaf forest, but there mostly occupy unfavourable sites such as ridge crests with thin and sometimes dry soils. Black beech can tolerate drier conditions than can hard beech. The montane beeches generally succeed conifer broadleaf forest altitudinally and sometimes they replace each other as dominants in a sequence of red, silver and mountain beech with increasing altitude. Both silver and mountain beech are more tolerant of coldness and soil infertility than red beech, but mountain beech is more tolerant of drought and so dominates drier Canterbury forests. The beech species all have small wind-pollinated flowers, but the stamens of male flowers may be attractively red.[96]

Bud protection

Many trees of regions with harsh winters, as well as losing their leaves, also protect the embryonic leaves at the twig tips with a number of broad, closely overlapping, hard and dry bud scales. These form a sealed case insulating against cold and drought. As the shoots expand in the spring the bud scales drop off leaving close-set groups of scars.

Species with such specialised bud protection are not common in New Zealand forests. Good examples are the species of southern beech.[96] The tree rata and pohutukawa, wineberry, and species of *Pittosporum*[97] also have overlapping bud scales. Some trees have smaller, narrower bud scales which may not be very closely overlapping, for example totara among the conifers, mahoe and the tree fuchsia; and others have a few fleshy scales, for example karaka.

At the other extreme are cases where there appears to be no bud protection at all; the immature leaves are completely exposed at twig tips. Here are such trees as the titoki, kohekohe,[98] puriri and tawa. Such unprotected buds are quite common in tropical forests and the preceding and other New Zealand examples have many tropical relatives.

However, even in the moist tropics there is some danger of young leaves drying out so they may have a covering of hairs later lost (New Zealand examples of this are rewarewa and rangiora[99]), or they may be protected by the sheathing bases or basal small appendages (stipules) of the mature leaves. In New Zealand sheath protection can be seen in kawakawa and five-finger among others and examples of stipule protection in species of *Coprosma* and in kamahi. Growing tips may also be kept moist by gummy (five-finger[100]) or mucilaginous secretions (*Coprosma*).

Once again the New Zealand pattern falls somewhere between those of tropical and strongly seasonal temperate forests.

98. (Opposite) Unprotected young leaves of kohekohe (*Dysoxylum spectabile*).

99. (Page 94) Young leaves of rangiora (*Brachyglottis repanda*) insulated with a close covering of hairs.

100. (Page 95) Unfolding young leaves, like hands held aloft, of five-finger (*Pseudopanax arboreus*). The young leaves are protected from drying out by a gummy fluid.

5. Juvenile and adult forms of trees

In the majority of trees there is no noticeable difference between the branching patterns and leaf sizes and shapes of young and adult plants. The leaves of the young plants are often shaded and at most tend to be somewhat larger and thinner than the adult leaves that are exposed to sun and wind.

About 90 per cent of New Zealand trees do not have a distinct juvenile form. The remaining 10 per cent however have juvenile forms, differing markedly from the adults in branching patterns and/or leaf sizes and shapes, that persist for a number of years. In extreme cases, such as the lancewood, specimens of the two phases collected on Cook's first voyage appeared so different that they were described as different species!

Distinct juvenile forms are not common in most temperate forests, but are more frequent in tropical forests. In the latter, juveniles of some trees may have: broad leaves of a similar shape to but much larger than those of the adults; compound leaves where the leaves of the adults are simple; or toothed, lobed or incised leaf margins where those of the adults are smooth (entire).

Large, broad juvenile leaves are rare in New Zealand. The common wineberry probably qualifies, but two other species are very localised in their occurrence—a shrubby coastal maire in the far north (*Nestegis apetala*) and the large-trunked *Coprosma chathamica* restricted to the Chatham Islands.

Juvenile compound leaves are more frequently encountered. The five-finger *(Pseudopanax)* relatives raukawa and haumakaroa[112] have juvenile leaves with leaflets arranged palmately and attached to a joint at the end of the leaf stalk. There is a transition after a few years to simple adult leaves. The very small-leaved shrubs *Pseudopanax anomalus*[110] and *Melicope simplex*[172] also have simple adult leaves and palmately compound seedling leaves. The common kamahi mostly has juvenile leaves with three leaflets, reducing to simple leaves in the adult. The far northern relative towai has up to five pairs of leaflets arranged in two rows with mostly three leaflets at the adult stage. In all these cases where the adult leaves are simple there is a joint at the end of the leaf stalk as an indicator of their compound derivation.

As in the tropics several New Zealand trees with smooth-margined or shallowly toothed adult leaves or leaflets have juvenile leaves that are lobed, sharply toothed or incised. In raukawa and haumakaroa the juvenile leaflets are often deeply and strikingly incised. The margins of adult leaves of the former are smooth and of the latter toothed. Titoki and pate also have incised juvenile leaflets, but in the case of pate, only in the northern North Island. Among woody vines *Clematis paniculata* has incised juvenile leaflets. Kohekohe and puriri have smooth-margined adult leaflets, but, respectively, lobed and toothed juvenile

leaflets. Several species have much more strongly toothed juvenile than adult leaves, for example wineberry and ngaio.

With regard to reduction in leaf size from juvenile to adult there is in New Zealand a special version probably not represented in tropical forests. Here the adult leaves are small, closely overlapping scales and, although the juvenile leaves are distinctly larger, they are still very small. The conifers are prominent here. In rimu the juveniles have strongly weeping, slender branches with short, needle-like leaves angled away from the twigs. There is a gradual change to the less weeping adult state with scale leaves. The juveniles of kahikatea branch rather untidily and their leaves are short spreading needles flattened into two rows.

102. (Opposite) A group of juvenile lancewoods (*Pseudopanax crassifolius*) with their unbranched slender trunks and terminal clusters of long, narrow, deflexed leaves. These were likened by a visiting botanical artist last century to 'collapsed umbrellas'. Standing above the uppermost juvenile is a young adult lancewood, much branched and with short, broad, upwardly directed leaves. Nga Manu Sanctuary, Waikanae.

103. (Above) The broadly rounded crown of a fully adult lancewood. To the right of it is a semi-juvenile which is beginning to branch and to produce leaves of intermediate form. Nga Manu Sanctuary, Waikanae.

104. (Below) Juvenile foliage and an adult leaf of the climber *Parsonsia heterophylla*. Juvenile foliage may be mottled brown as here and first-formed leaves are usually round and later ones long and narrow.

Again the transition to the adult state is fairly gradual. By contrast the rather longer needle-like leaves of the juveniles of monoao, silver pine, yellow-silver pine and others[101] give way abruptly to the adult scale leaves.

Another type of juvenile/adult contrast in New Zealand, particularly striking in some examples, involves long, narrow, sometimes deflexed juvenile leaves at the top of slender unbranched stems. The adults are freely branched with much shorter upwardly directed leaves. The most striking example of this, and the best known, is the lancewood. Here the stem of the juvenile, although very tough and woody, remains slender and unbranched until it reaches a height of six metres or more after about 15 years.[102] Hanging down from the top of this slender trunk are a cluster of long, narrow, sharply toothed leaves that can be a metre or more in length, but only a centimetre or less in width. Eventually the top of the trunk branches freely to form a broadly rounded crown bearing adult leaves that are usually no more than 10 cm long by 2-3 cm wide and upwardly directed.[102,103] Lancewood to most people means the juvenile stage, and along with Captain Cook's botanists they would not realise that the adult phase was the same plant. The fact that they are the same is dramatically demonstrated when an adult lancewood forms a branch from the base of its trunk. This is always completely juvenile in form and alien in appearance to the rest of the tree. The formerly dormant bud that grows into such a branch would have been formed when the tree was juvenile and so would have retained that 'programme'. A much less common species related to the lancewood (*Pseudopanax ferox*) has the same basic life cycle, but here the deflexed juvenile leaves have quite large, jagged, hook-like lobes along their margins.[105] It has been suggested that the toughness of the juvenile leaves of these two species, along with their toothed or jagged margins, would have discouraged moas from browsing on them. Another idea is that the growth effort of these juveniles is concentrated on increase in height and it is only when competing plants are overtopped, particularly in second growth forest, that a spreading crown is formed.

In other species with narrow-leaved juveniles the juvenile/adult contrast is not so dramatic. Although the juveniles of rewarewa and hinau have narrow leaves, the branching of their stems is not so long delayed. The juvenile leaves of these two species are sometimes difficult to tell apart, but those of rewarewa generally have a stiffer texture and coarser teeth.

The species of maire, excepting *Nestegis apetala*, also have juvenile leaves that are narrower and sometimes longer than the adults.

As remarkable as the lancewoods, but with a very different pattern, are the trees with very small juvenile leaves borne on slender interlaced twigs. Probably the commonest of these is the small tree kaikomako.[106,113] The juvenile has an abundance of slender, wiry, widely-angled (divaricate) twigs, that become so interlaced that they form a ball-like springy mass. The leaves are usually less than a

101

centimetre long and wide with three, pointed lobes. It is a number of years before the adult state begins to form at the top of the twiggy mass. The adult branches are fewer and more narrowly angled and the adult leaves are many times larger than the juveniles with more teeth or lobes. After an adult crown has developed the juvenile phase may persist for some time towards the base of the trunk, but eventually it dies away. Putaputaweta has a similar divaricately branched and small-leaved juvenile, but the leaves are somewhat larger and the branching is more orderly with the often zig-zag lateral branches arising in tiers of three or four on the main axis.

Ribbonwood, lacebark and its relatives also have juveniles with small, often deeply lobed, leaves and range from strongly to moderately divaricate.

The common kowhai (*Sophora microphylla*) at some localities has a very untidy, tortuous and small-leaved juvenile. This may persist for as long as thirty years before a more orderly adult branching pattern develops with larger leaves.

106. A kaikomako (*Pennantia corymbosa*) with excessively twiggy small-leaved juvenile growth below grading upwards into the more sparsely branched larger-leaved adult phase. Nga Manu Sanctuary, Waikanae.

Coprosma arborea and black and mountain beech also have small-leaved juveniles with moderately divaricate branching.

Standing apart in one respect, the small-leaved, very divaricate juvenile of pokaka has leaves remarkably variable in size and shape.[114] They may be variously incised or toothed and range from long and narrow to completely round with every combination between. The juvenile leaves of *Pittosporum turneri* are similar, as are those of the woody vine *Parsonsia heterophylla*.[104]

The only conifer to have a divaricating juvenile is matai. The widely spreading entangled twigs are often downwardly curved and they and the often sparse leaves are often coloured light-brown, contrasting with the dark-green of the adult foliage.

Long and narrow juvenile leaves and small-leaved divaricate juveniles are not to be found in most tropical forests, but it is interesting to note that on the dry sides of the islands of the Seychelles in the Indian Ocean a number of trees have small-leaved juveniles. Some of the latter have very narrow leaves and in others they are rounded. None appears to have divaricate branching however.

6. Moa, Ice Ages and small-leaved shrubs

At the end of the last chapter we considered examples of New Zealand trees with twiggy, small-leaved juveniles. There are at least seven such cases, but this is just the tip of the iceberg as there are in addition about 55 shrub species of this form. These are generally termed 'divaricating shrubs' in reference to their wide branching angles. Altogether about 10 per cent of the woody species in New Zealand are divaricating shrubs, or trees with divaricating juveniles.

As well as having a number of divaricate species, some larger genera also have species with leaves sometimes hundreds of times larger than those of the divaricates and with narrow branching angles. Examples are *Coprosma*, *Pittosporum*, *Melicytus* and *Olearia*. In *Pseudopanax*,[112] *Myrsine*[107,108] and *Muehlenbeckia* only one species has the divaricate form. Several other genera where this habit is exhibited are represented in New Zealand by only one or two species.

Dense stands of this growth form, so interlaced within and between individuals that a person's weight can be supported,[109] have impressed and puzzled both resident and visiting botanists for more than a century. Theories have proliferated and are still being debated actively today. To the untutored eye these divaricates all look the same and if there were in fact only one species it could be dismissed as a chance

107. *Myrsine divaricata* in a relatively sheltered site with its typical downwardly curved twigs. Cobb Valley, north-west Nelson.

aberrant form that happened to find a suitable niche. The fact that there are so many different species of this form, most of them quite unrelated, suggests that some common environmental factor or factors has selected for this specialisation. What that factor might be is the matter in dispute.

The first suggestion, last century, was that it was an adaptation to drought as small leaves in desert plants are considered to be a means of reducing the amount of water lost by evaporation. The small-leaved divaricate form in New Zealand would probably have evolved from forest ancestors during glacial periods of the recent Ice Age when climates would have been drier, particularly in the rain shadow to the east of the axial mountains. Arising from the fact that divaricate juveniles are particularly abundant in forest patches in eastern districts, it has also been suggested that such juveniles are suited to present day dryish forest climates. The limited amount of water absorbed by their shallow root systems is conserved by the small leaves, but once a deeper root system has developed moister soil levels can

be tapped and larger, more efficient adult leaves can be supported. In the even drier conditions during glacials, with the added effect of soil coldness reducing the absorptive efficiency of the roots, these small-leaved juveniles could become fixed as permanent divaricating shrubs.

More recently the 'climate theory' has been revisited and it has been concluded that the small-leaved divaricating habit is well adapted to windy, frosty and drought-prone habitats in New Zealand. It is pointed out that New Zealand's narrow and isolated land mass has a very oceanic climate with short-term fluctuations rather than the longer-term regular seasonal patterns of continental areas. Frosty nights can be followed by quite warm days, even in winter, and violent gales in the 'roaring forties' can be quickly followed by calm sunny weather. The divaricates can withstand the inclement spells and quickly resume active growth in the more genial intervals. The proponents of this view note that in many divaricates the outer and youngest layer of twigs has very reduced leaves and that fully developed, but small, leaves are restricted to older growth in the interior.[110] They see the outer twiggy mass as providing a wind screen, allowing the retention of relatively moist air within, a frost screen, and perhaps on sunny days a heat trap. The fact that divaricates are particularly abundant on low-lying alluvial sites in drier eastern areas is explained as being the result of the frosti-ness of such sites. A similar explanation has been offered for the promi-nence of divaricates in early regeneration on frosty river flats on the much wetter west coast of the South Island. Where divaricate juveniles occur in frosty sites the changeover to the larger-leaved adult phase, they suggest, takes place above the level of damaging ground frosts.

108. (Opposite) *Myrsine divaricata* in a gale-swept situation with a short trunk, tortuous branches and a dense twiggy crown. Auckland Islands.

109. (Above) A wind-shorn shrubbery of a number of divaricating shrubs and juveniles on Mt Kaukau, Welling-ton. (Photo: John Dawson)

A quite different idea, formulated in detail fairly recently, holds that divaricates are browse-resistant forms that evolved in response to the presence of moa. Moa were large flightless birds that are considered to have substituted for land mammals in New Zealand. They became extinct only a few centuries ago. The proponents of the 'moa theory' accept that small-leaved divaricate shrubs can be found elsewhere in the world, for example at a dry coastal locality in the south of Madagascar; in Patagonia on the drier side of the Andes in southern South America; and, from my own observations, on dry hill slopes in Greece. However, in these places the majority of divaricates are very spiny, presumably to discourage browsing by soft-muzzled mammals. In New Zealand such spines would have been ineffective against hard-beaked moa and the great majority of our divaricates are without spines.

It is suggested: (1) that the often dead-looking outer twigs of divaricates would appear unappetising to moa;[110] (2) that the tough, wiry twigs would be difficult to pull off and then to disentangle; and (3) that the many, widely separated growing points of divaricates would ensure that enough of them would be left after browsing for the ready replacement of lost twigs and leaves. (The later discovery in the crops of moa preserved in swamps of twigs of divaricates with cleanly cut ends indicates that at least some species of moa did not have difficulty in detaching twigs.) According to the moa theory too, divaricates are common on fertile alluvial soils because plants growing in such sites would be more nutritious and therefore sought after by moa. With divaricate juveniles the changeover to the more appetising adult state would take place above the reach of moa.

Other features of New Zealand plants may also have deterred moa browsing: the excessively spiny spaniards of open, mostly mountain, habitats—here the spines are so long that they would have gone past the beaks of moa and threatened their eyes; plants with very tough leaves such as those of the cabbage tree, *Phormium* and the juveniles of lancewoods; forest trees and shrubs with poisonous or unpleasant-tasting leaves, for example, the peppery leaves of kawakawa and horopito; palatable species that look like (mimic) an unpalatable species—for example *Alseuosmia pusilla* is very difficult to distinguish from mountain horopito and often grows with it; camouflage colouring of young forest plants—the juveniles of several woody forest plants have brown leaves and are sometimes mottled with several shades of

110. (Above) *Pseudopanax anomalus* showing dead-looking recent growth above. Otari Native Botanic Garden, Wellington.

111. (Opposite) Young lancewoods (*Pseudopanax crassifolius*) whose mottled brown colouration merges with that of the beech litter. Lake Rotoiti, Nelson Lakes National Park.

brown from pale to dark and are very difficult to see against leaf litter or tree trunks. Examples are kahikatea, matai, lancewood,[111] pokaka[114] and the vine *Parsonsia heterophylla*.[106]

It is clear that the arguments for the moa and climate theories have been well researched and formulated, but it is clear also that the debate will continue for some time.

A number of divaricate species are able to grow and reproduce in forest shelter and shade; in some cases, perhaps, as a result of persisting from a more open phase. However, in many forests on slopes in moister climates the small trees and shrubs below the canopy are large-leaved and if divaricates are present at all they are at the more exposed margins or in larger canopy gaps. As already noted they can be common on alluvial terraces, in the open or under a light forest canopy. At higher altitudes they contribute to mountain shrubland and forest openings. They can also be found at forest margins, in swamps, along exposed coasts and on gale-swept hill tops near Cook Strait.

If this distinctive growth form was induced by moa browsing then it must be very ancient as moa are believed to have been in New Zealand for perhaps 80 million years. If, however, it was induced by Ice Age drought and cold then it must be a more recent phenomenon. The frequency with which certain pairs of species, one divaricate and one 'normal', hybridise in nature lends support to the latter view. Such hybridising takes place in *Coprosma* (several pairs of species), *Melicytus, Olearia, Myrsine, Corokia, Pseudopanax,*[112] *Melicope, Elaeocarpus,*[114] *Aristotelia* and *Plagianthus* as well as spontaneous garden hybrids in *Pennantia*.[113] In most cases the first generation hybrids are partly fertile and give rise to 'hybrid swarms' exhibiting every combination of the characteristics of the two parents. The parent species in each case, despite being very different in appearance and ecology, must still be fairly compatible genetically. This suggests that one parent, presumably the divaricate, has been derived from the other, or that both have had a common ancestry, in recent geological times.

It has been suggested that plants that have been selected in nature for hardiness in respect of one limiting factor in the environment may also prove to be hardy in respect of other factors. With small-leaved divaricates throughout the world there seems to be a linking thread of drought tolerance. In areas with browsing mammals there is the added feature of spininess. In New Zealand, therefore, this may be a basically drought-tolerant form, which is also wind resistant, frost hardy and browse resistant. Alternatively, the basic adaptation of this form may relate to browse resistance, against mammals in most parts of the world and moa in New Zealand. In this case it would be the climatic tolerances that would be secondary.

At the very least, we will end up being much better informed, if not more enlightened, about this phenomenon and related matters as a result of the controversy.

112. *Pseudopanax.* On the right the compound, dissected juvenile and simple but toothed adult leaves of haumakaroa (*P. simplex*), on the left the twiggy small-leaved *P. anomalus* below, and a portion of a natural hybrid between the two species above. Lake Rotoiti, Nelson Lakes National Park.

113. *Pennantia.* At the centre juvenile (above) and adult foliage of kaikomako (*P. corymbosa*). On the left a leaf of *P. baylisiana* from the Three Kings Islands off the northern tip of the North Island. At right foliage of the garden hybrid between the two. When discovered on the Three Kings, *P. baylisiana* had been reduced to a single female tree. This was propagated from cuttings and a tree is now growing and flowering at Otari Native Botanic Garden in Wellington. This plant sets some berries as a result of pollination by local *P. corymbosa* and vigorous hybrids are now in cultivation. It has been found that female flowers of *P. baylisiana* occasionally produce a little pollen and a few new plants of the species have now been obtained from seed.

114. *Elaeocarpus.* On the upper left, twiggy, variable-leaved juvenile foliage of pokaka (*E. hookerianus*). On the right, long-leaved juvenile leaves of hinau (*E. dentatus*). At lower left, the juvenile of a naturally occurring hybrid between the two. Kaitoke Regional Park, Upper Hutt.

7. Climbing to the light—the vines

Vines, epiphytes and mistletoes—What's the difference?

All three depend on the physical support of free-standing trees or shrubs to gain adequately lit sites for their foliage above the shade of the forest floor.

The seeds or spores of vines germinate on the ground and the stems of the young plants climb shrub stems or tree trunks by a variety of means.

The seeds or spores of epiphytes germinate on the branches or sometimes the trunks of trees. Herbaceous epiphytes and small shrub species never have root contact with the ground, while larger shrub and tree epiphytes eventually do.

The seeds of some parasites (mistletoes) also germinate on tree or shrub branches, but whereas epiphytes just attach to the surface of their supports, parasites have special root-like organs that penetrate the inner tissues of their hosts and draw from them water, mineral nutrients and sometimes organic nutrients.

Subcanopy vines

In New Zealand these are all ferns, and, although some can attain considerable heights above the ground, they mostly do not extend into full light of the forest roof. They climb by firmly attaching their

115. (Opposite) Vines of the fern *Blechnum filiforme* growing up tawa trunks. On the left is a young plant with the small juvenile leaves, on the right an older plant with large adult leaves at the top intermingled with a few fertile leaves with their string-like segments.

116. (Below left) The climbing fern *Phymatosorus scandens*: narrow, undivided juvenile leaves near the ground; larger adult leaves above divided into leaflets.

117. (Below right) The stout, flecked climbing stems of *Phymatosorus diversifolius*. Wellington Botanic Garden.

upwardly growing stems to tree trunks with special short roots. In some cases these ferns first spread over the forest floor, forming quite a carpet of fronds, before climbing trunks they encounter. As they climb several behave in an interesting and puzzling fashion: each leaf formed is larger than the one below it until a maximum size is attained. Thus in *Blechnum filiforme*,[115,118] leaves on the forest floor or lower tree trunks are no more than ten cm long with two rows of small leaflets. There is a steady increase in size upwards until at the maximum one leaflet, now having an elongated shape, is about as long as a complete juvenile leaf. It is only among the largest upper leaves that the fertile fronds with their almost thread-like leaflets are formed.

The thin-leaved *Phymatosorus scandens*[116,118] exhibits a similar pattern, but here the leaves at lower levels are long and narrow and are not divided into leaflets, while those higher up are compound with two rows of long narrow leaflets. The thick-leaved *Phymatosorus diversifolius*, with its stout scale-flecked stems,[117] also has a range from simple to compound leaves, but, rather than vertical trunks, this species seems to be most abundant on trees with inclined, spreading, well-lit branches. *Arthropteris tenella* is a climbing fern of warmer forests often near the sea. The slender stems are very scaly and the juvenile leaves are only a few cm long, sometimes with a few pairs of small rounded leaflets and a disproportionately large terminal leaflet. Adult leaves are up to 30 cm long, bearing many pairs of elongated leaflets with attractively scalloped margins.

There seems no obvious explanation for the increase in size and change in shape of the leaves with increasing height above the ground. Some subcanopy vines in tropical forests behave in a similar fashion: the quite unrelated aroids (arum lily family) of these forests also have small and often simple leaves near the ground and much larger and often compound leaves higher up.

Canopy or sun vines

These are more numerous than subcanopy vines and many of them at maturity have stout, woody, cable-like stems similar to those used by Tarzan. Others have more slender stems that are, nevertheless, hard in texture and wiry. The devices for climbing are various.

(a) Attaching roots

Prominent here are the several species of climbing rata (*Metrosideros*). Several of them, including the white climbing rata (*M. perforata*),[119,120] and the red climbing rata (*M. fulgens*), do not form leafy, flower-bearing branches, extending away from their supports, until the climbing stems have reached the tree crowns. These climbing stems have quite short attaching roots. In *M. diffusa* and *M. colensoi* the flowering non-climbing branches, attractively curtain-like in the latter, often form at quite a low level on tree trunks. In these cases the climbing stems are often attached by longer roots that sometimes go right round slender tree trunks.

118. (Page 112) Two climbing ferns with juvenile foliage, on the same tree trunk: *Blechnum filiforme* on the left, *Phymatosorus scandens* on the right. Note the dense ground cover of the introduced weed *Tradescantia*. Battle Hill Reserve, Wellington.

119. (Page 113) The leaves of a young white climbing rata (*Metrosideros perforata*) forming a mosaic on a rimu trunk.

120. (Opposite) Massive woody cables of two adult white climbing rata (*Metrosideros perforata*) swinging away from their supporting rimu. There is also a younger vine forming a leaf mosaic on the trunk. Kaitoke Regional Park, Upper Hutt.

Juvenile plants of the climbing ratas are very similar to each other with their small round leaves. The leaves increase in size and often change in shape with increasing height above the ground. (Perhaps there is a general tendency for root climbing vines to have small-leaved juveniles).

Three of the climbing ratas occur together through most of the country: *Metrosideros fulgens* has largish leaves and large, handsome red to orange flowers[122,132] from autumn through winter and into spring; *M. perforata*[121] and *M. diffusa* have small white or pinkish-white flowers in summer. *M. colensoi*, also with small white flowers in summer, is more localised being largely restricted to fertile sites such as river terraces and limestone outcrops. *M. carminea* and *M. albiflora* are restricted to the north of the North Island, although the former is now widely cultivated for its striking display of red flowers in early spring.[123] *M. albiflora* has the largest leaves of the climbing ratas and small but often abundant pure white flowers in summer. It is often to be found in kauri forests.

Our only other canopy vine with attaching roots is the kiekie, a very different plant. Kiekie belongs to a widespread tropical genus (*Freycinetia*) and has long, narrow, dark green leaves with finely toothed cutting edges. Its tough stems, up to 4 cm thick, are attached to tree trunks by slender horizontal roots that sometimes extend right round the trunk.[124] These attaching roots are not often seen as they are soon obscured by a profuse growth of a second thicker type of root that grows down the trunk to the ground.[125,168] Kiekie seems to be particularly abundant in swampy forest sites where it covers the ground in open places and elsewhere completely obscures trunks, such as those of kahikatea and pukatea, with its stems, descending roots and foliage.

121. (Page 116) The small white flowers of the white climbing rata (*Metrosideros perforata*).

122. (Page 117) Handsome buds and flowers of the red climbing rata (*Metrosideros fulgens*). This species has an unusually long flowering period from February to September. Otari Native Botanic Garden, Wellington.

123. (Above) *Metrosideros carminea*, a vine of the northern North Island now widely cultivated for its attractive flowers in early spring. Otari Native Botanic Garden, Wellington.

124. (Opposite) Horizontal anchoring roots of kiekie (*Freycinetia banksii*).

118

The upper parts of the stems often separate from the tree trunks and loop downwards, adding a distinctly tropical, if somewhat untidy, aspect to the scene.[60,125]

(b) Climbing with hooks

The vines in New Zealand climbing by hooks are our several species of 'lawyer'. The lawyers belong to the same genus, *Rubus*, as the introduced blackberry and raspberry.

Young lawyer vines have slender but tough stems that grow upward strongly and if they grow through the branches of a shrub or small tree cling to them firmly with the tough recurved hooks on their stems, leaf petioles (stalks) and leaflet midribs.[127] Once firmly anchored, hooked branches can extend even higher by the same means until well-lit tree crowns are reached. By this stage the main stem extending up from the ground has become large and woody. Sometimes it is partly looped on the forest floor and can give rise to new leafy shoots.

The common bush lawyer (*R. cissoides*) is the largest species. It has narrow, shiny leaflets that are quite glabrous (without hairs). *R. schmidelioides* has leaflets of a similar shape, but much smaller, that often appear grey below from a covering of short hairs. The swamp lawyer (*R. australis*) has small, quite round leaflets. *R. squarrosus* is remarkable when growing in the open. It forms tangled masses and appears to be leafless. However, there are leaves but with their leaflets reduced to the midribs. All the axes are liberally provided with yellow hooks—a daunting sight.[126] When this species climbs up into a tree, leaflets with narrow blades are formed.

125. (Page 120) A profusion of looping stems and long narrow leaves of kiekie on a buttressed pukatea. Masses of thick descending roots from the kiekie can be seen against the trunk. Nga Manu Sanctuary, Waikanae.

126. (Page 121) The slender stems and leaf axes of *Rubus squarrosus* ready to cling to any support with its many small hooks.

127. (Above) Young leaves of *Rubus schmidelioides* beset with recurved hooks.

(c) Twining vines

Several vine species use their stems as climbing devices. As stems of young plants grow upward from the forest floor their tips slowly rotate, as can be dramatically demonstrated with time lapse photography. In some species this rotation is clockwise, in others anticlockwise. Only rarely is there a change from one direction to another. If a sufficiently slender support is encountered the vine twines steadily upward around it. Both support and vine stems increase in diameter with time and often the vine can become embedded in the support.[128] Sometimes the support stem dies within the coils of the vine, but by this time the upper branches of the vine may already have become established in crown of a tree. Sometimes too the vine stem may be stretched and broken as the support stem expands and in this case the support retains a corkscrew form as evidence of the encounter.[129]

We will first consider two slender twining vines, one a fern and the other a fern ally (*Lycopodium).*

The fern is mangemange (*Lygodium articulatum). Lygodium* is a widespread tropical genus and our species is restricted to the north of the North Island. Strictly speaking, what appear to be twining stems here are the axes of leaves with indefinite growth. The true stems are quite short on the forest floor. The wiry 'leaf stems' bear compound leaflets some of which form spore-producing organs. As well as twining about supports the axes twine about each other to form springy entanglements, in open places on the ground or on tree branches.

128. (Above left) A twining stem of *Parsonsia heterophylla* partly embedded in a kohekohe branch.

129. (Above right) A young matai (*Prumnopitys taxifolia*) with a corkscrew trunk as a result of an earlier encounter with a twining vine. (Photo: John Dawson)

Lycopodium volubile is an attractive, usually trailing plant with delicate spreading fern-like branchlets. The stems bearing the small green leaves spread over the ground in openings or scramble over shrubs, but it has been noted that stems bearing the distinctive drooping spore-forming cones twine around the stems of shrubs or small trees.[130] In New Zealand this species does not appear to climb particularly high, but it also grows in the Asian tropics and there it is described as a high climbing vine.

There are also several more substantial twining vines.

Supplejack is a very familiar and widespread species which sometimes forms almost impenetrable tangles on the forest floor.[131] The twining stems never become more than about 2 cm in diameter, but they are very hard and cane-like and almost black in colour. The leaves, reduced to long twisted scales, are also black and the tips of the shoots are soft and resemble asparagus spears. Once a stem has reached the forest roof others may twine around it until quite massive multi-stranded cables are formed.[132] In these cases the original supporting stem would have long since died. In the better light of tree crowns a second and quite different type of branch is formed. It is slender, up to about a metre long and bears two rows of broad, green, functional leaves. These stems never twine and only they form the small greenish flowers[133] and later the bright red berries.[134]

Parsonsia heterophylla, sometimes known as New Zealand jasmine, is also widespread, but here the stems become quite woody[128] and up to 10 cm in diameter. The stem surface becomes rough with small, blunt outgrowths. The leaves of juvenile plants are very small and very variable in shape.[106] Usually the first leaves on a branchlet are completely round and the last formed long and narrow with strangely shaped intermediates in between. Adult leaves are larger and broader and uniform for each plant.[106] The clusters of small white flowers in spring are quite like those of the jasmine and are pleasantly perfumed. The long pod-like capsules split open to release small seeds with parachutes of hairs.

The common *Muehlenbeckia australis* also has woody stems about 10 cm in diameter and broad but quite thin leaves. Leaves on young plants are smaller and some of them may be distinctively fiddle-shaped. This vine is often particularly abundant in second growth forest where its foliage sometimes smothers that of the supporting trees.

Two other woody vines in New Zealand have different methods for climbing.

In *Clematis paniculata*, notable for its festoons of large, pure white, star-like flowers in spring,[135] it is the stalks of the leaves and leaflets that twine. These stalks are touch sensitive and twine around any slender stem or other axis with which they come into contact. Larger stems are not suitable, so the *Clematis* has to attain the forest roof by climbing via shrubs and smaller trees particularly at forest margins. Adult leaves have three broad leaflets, but in young plants the first leaves are simple,

130. (Page 124) The climbing *Lycopodium volubile*, with attractive spreading foliage below and twining stems with spore producing cones on manuka stems above. Piupiu Springs, Takaka.

131. (Page 125) An entanglement of cane-like supplejack stems (*Ripogonum scandens*). Otari Native Botanic Garden, Wellington.

132. (Opposite) On the right are flowers of the red climbing rata (*Metrosideros fulgens*). On the left is a multi-stranded cable of the twining vine supplejack (*Ripogonum scandens*). Wainui Waterworks Reserve.

long and narrow, and soft in texture. Subsequent leaves have three narrow rather broader leaflets variously lobed and incised. The adult leaflets have smooth margins.

There are several smaller species of *Clematis* that mostly grow on shrubs in the open or in low second-growth forest.

The native passion vine, *Passiflora tetrandra,* has twining tendrils. In this case they are considered to be modified branchlets and arise just above where the leaves attach to the stems. Like the leaf stalks of *Clematis* these tendrils are sensitive to touch and quickly wind round any slender support. When they have firmly attached, the free parts of the tendrils coil up like springs drawing the passion vine stem closer to the support. Eventually the stems of the vine become enlarged and woody and their weight can cause them to partly collapse to the forest floor where they form snake-like coils. Compared with their cultivated relatives the flowers of our species, although still quite elaborate in form, are small and lacking in colour. The fruits, however, are very attractive, appearing like small, bright orange balloons against the dark green foliage.[136]

133. (Below left) The strange, small flowers of supplejack. What appear to be 6 plump petals are actually pollen-containing anthers.

134. (Below right) Leaves with their distinctive vein patterns and berries of supplejack.

135. (Above) Pure white female flowers left and male flowers right of *Clematis paniculata*. The male flowers have attractive pink anthers on their stamens.

136. (Below) The clear orange fruits of the native passion vine (*Passiflora tetrandra*) stand out against the dark green foliage.

8. Epiphytes and mistletoes

Most moist temperate forests have only small lichens, mosses and liverworts as epiphytes on trunks and branches. These are common in New Zealand as well and, here as elsewhere, they are particularly abundant in higher altitude forests that are often shrouded in mist. Large lichen and moss cushions may be abundant on trunks and major branches and more delicate hanging mosses and lichens, sometimes known as 'old man's beard', sway in the wind from every twig.[137] This type of forest is known prosaically as cloud forest or mossy forest or, more romantically, as 'goblin forest' or 'elfin woodland'. Some very small lichens, leafy liverworts[138] and algae also grow closely pressed to the surfaces of leaves of some New Zealand trees and vines, although this type of epiphyte is much more conspicuous on the larger leaves of the tropics.

In tropical forests the conspicuous and distinctive epiphytes are larger vascular plants (ferns and flowering plants), and although New Zealand is a temperate country this is also true of our conifer broadleaf forests. Vascular epiphytes that are known as sun epiphytes require a high level of light to grow and reproduce; they gain this from the outset by establishing in the branches of trees. What they don't have is the reliable water and mineral nutrient supply that most other plants obtain from the soil. This lack prevails through the entire life cycles of smaller epiphytes and for many years in the cases of larger ones.

The specialisations of vines enable them to climb to the light; those of epiphytes enable them to obtain and conserve water that is often in short supply in tree tops as a result of drying winds, lack of soil and rapid drainage.

We will first consider the epiphytes of ordinary trees, then the special case of epiphytes of tree fern trunks.

Shade epiphytes

The shade epiphytes attach to the lower parts of tree trunks where they escape the deep shade and competition of forest floor plants.

Several small, tufted ferns, attached by roots, are prominent in this situation. All of them have brown to orange, dome-like sporangia on the leaf undersides. The three species of *Grammitis* and *Anarthropteris lanceolata* have narrow simple leaves, while those of *Ctenopteris heterophylla* are twice compound. The *Anarthropteris* alone has the ability to form new tufts of leaves from its spreading roots.

A much larger fern is *Phymatosorus novae-zelandiae* in higher altitude forests of the North Island. Its quite large leaves, divided into narrow leaflets, are attached to short stout stems covered with straw-coloured scales.

137. (Previous two pages) Mountain beech forest on the Lewis Pass draped with old man's beard lichen.

138. (Opposite) A moss and a leafy liverwort growing on the leaf of a nikau palm.

Two small species of the large succulent genus *Peperomia* also grow as low epiphytes, one in the north of the North Island and the other extending to the north of the South Island.

All these low epiphytes can be found in moderately moist to dryish forests, but a number of additional fern species establish on tree trunks and branches in wet forests at higher altitudes and on the wetter western sides of both islands. Trunks and branches in these conditions support coverings of lichens and moss cushions as well as up to ten species of delicate filmy ferns (*Hymenophyllum*). Some of them are very small and moss-like; others are larger and hang down from the trunks.

Sun epiphytes

These are much more diverse than the shade epiphytes and can be grouped into herbaceous and woody species.

(a) Herbaceous epiphytes

The most conspicuous here are the nest epiphytes[139,140] comprising three species of the lily family in the wide sense—*Astelia solandri*, *Collospermum hastatum* and *C. microspermum*. These all have very short, branching stems bearing dense tufts of long, narrow leaves above and long attaching roots below. When the crown of a tall tree supports many of these nests on its branches, each a metre or more across, they certainly do look like exceptionally large birds' nests. Over time, deep, black humus builds up below the nests from the decay of their own old leaves and roots with a contribution from bark flakes and leaves of the supporting tree. This humus provides nutrients to the epiphytes, but, more importantly, it acts as a very efficient sponge to absorb and retain rain water. The silvery-green leaves of *Astelia* are narrow and closely overlapping at the base, but those of the collospermums, arranged in wide fans,[141] are very broad and rounded at the base with enclosed spaces or 'tanks' between that can store significant amounts of water after rain.[142] Branching roots grow between dying outer leaves to absorb this water, and it has been suggested that the inner leaves can absorb it directly through special gland-like structures, although this has not been proved. Similar glands are known to absorb water in the unrelated tank epiphytes (bromeliads) of the Amazon forests.

Nest epiphytes, with their associated humus masses, eventually become very heavy, especially after rain, and they often fall to the ground where they usually die from lack of light. During rain with strong winds the weight of epiphyte nests can sometimes cause whole branches to break off and crash to the ground.

Other herbaceous epiphytes hang below the epiphyte nests with their roots penetrating into the humus. The fern relative *Lycopodium varium* forms heavy, tassel-like masses of freely branching, small-leaved stems that are set in motion by the lightest breeze.[143] The long fronds of the ferns *Asplenium polyodon*[144] and *A. flaccidum*[143] are also frequently seen below nests, the former with attractively toothed rhomboid leaflets, the latter with long, almost string-like leaflets.

134

139. (Opposite) An epiphyte garden. The plants are perched on the trunk and branches of a northern rata (*Metrosideros robusta*), that also started its life as an epiphyte. In order downwards from the upper left: a nest of the light-demanding *Collospermum hastatum* with its fans of leaves; a nest of the more shade-tolerant *Astelia solandri* with drooping leaves; the fern *Asplenium flaccidum*; and, lastly, foliage of the small shrub epiphyte, *Pittosporum cornifolium*. To the right there are other nests of *Collospermum* and *Astelia*, some with fronds of the larger epiphyte fern, *Asplenium polyodon*, hanging below them. Wainui Waterworks Reserve, Wellington.

140. (Page 136) A tall kahikatea with many nest epiphytes, mostly of the more light-demanding *Collospermum hastatum*. Wainui Waterworks Reserve, Wellington.

141. (Previous page) A leaf fan of the nest epiphyte *Collospermum hastatum* with characteristic dark bases.

142. (Left) View of a *Collospermum* leaf fan from above showing water glinting in the reservoirs between the broad leaf bases.

143. (Opposite) The epiphytic *Lycopodium varium* which often forms heavy tassels of forking stems below nest epiphytes. The large shiny leaves below belong to a young plant of the large epiphytic shrub puka (*Griselinia lucida*). At upper left is the epiphytic fern *Asplenium flaccidum*.

Other herbaceous epiphytes attach directly to trunks and branches, sometimes in the thin soil formed by mosses and lichens. These are mostly orchids. The largest is *Dendrobium cunninghamii*, whose hanging, freely branching stems and small leaves have a billowing appearance. The polished stems have been likened to slender bamboo with their pronounced joints. The flowers are up to two and a half cm across, very modest compared with some of their tropical relatives, and are white with a rose-pink centre.[146] Two earinas have narrow leaves like the *Dendrobium*, but their usually hanging stems do not branch. The leaves are in two rows, slender and grassy in *Earina mucronata*,[145] more robust in *E. autumnalis*. Both have clusters of small flowers with a spicy perfume, those of the former appearing in spring and the latter in autumn.

The two species of *Bulbophyllum* are very different. Their leaves and flowers are very small and their slender stems branch and spread to form leafy patches. Each leaf is swollen at the base into a water-storing pseudobulb.

Drymoanthus adversus also has very small flowers, but longer tufted leaves. The short stems do not branch and give rise to widely spreading web-like roots. This species attaches to trunks as well as branches.

The orchids all have roots with an outer zone of dead empty cells that absorb and retain water.

One fern, *Pyrrosia eleagnifolia*, forms extensive patches with its spreading slender stems over well-lit trunks and branches.[147] The short, rounded simple leaves are quite fleshy with water-storing tissue and are covered with scaly hairs that would help to reduce water loss.

(b) Woody epiphytes

These range from small shrubs to large trees.

Among the small shrubs, all establishing in nest epiphytes, are two species of *Pittosporum*. *Pittosporum cornifolium* extends to the northern South Island and is common as an epiphyte high in tree crowns. Its roots anchor into the humus below nest epiphytes and its rather slender, spindly stems generally droop downwards.[148] The seed capsule of this species is spectacular when it opens in late autumn and winter. The linings of the two valves are brick red and seated at the centre is a cluster of shiny black seeds embedded in a sticky bright yellow fluid.[149,150] Birds are attracted by the bright colours, and the seeds stick to their beaks to be transported to other trees. *Pittosporum kirkii* is restricted to the northern half of the North Island. It has a more upright habit with somewhat fleshy water-storing leaves. The flattened capsules, although quite large, are not as colourful when open as those of *P. cornifolium*.

A daisy shrub, *Brachyglottis kirkii*, is a common epiphyte in the North Island. It has somewhat fleshy, toothed leaves and broad heads of pure white flowers.

The shiny karamu, *Coprosma lucida*, although better known as a ground plant in open forest or shrub, is also quite common as an epiphyte.

Outstanding among large shrub epiphytes is the puka. Its rounded crowns of large, shiny leaves, contrasting with the usually much smaller leaves of the supporting trees,[151] are frequent in conifer broadleaf forest through much of the country. This species, too, often establishes in a nest epiphyte,[143] but eventually sends a slender root down to the ground. This root and branches from it grow closely pressed to the bark of the supporting tree or even behind bark flakes in the case of the rimu.[152] Once the root reaches the ground and a good supply of water and mineral nutrients, growth of the whole plant speeds up and in particular the descending root enlarges greatly and becomes deeply and characteristically grooved longitudinally.[154,155]

144. (Page 140) The epiphytic fern *Asplenium polyodon* with its fronds hanging below the nest epiphyte *Collospermum microspermum*. Near Erua, central North Island.

145. (Page 141) The epiphytic orchid *Earina mucronata* with small but attractive flowers.

146. (Opposite above) A group of flowers of the epiphytic orchid *Dendrobium cunninghamii*.

147. (Opposite below) The fleshy-leaved epiphytic fern *Pyrrosia eleagnifolia*, forming a mat-like cover on a branch. (Photo: John Dawson)

148. (Opposite) The slender drooping stems of the small shrub epiphyte *Pittosporum cornifolium* growing out from a nest epiphyte.

149. (Above left) The surprisingly colourful open capsules of the epiphyte *Pittosporum cornifolium*. The inner lining of the capsules is brick red, the seeds are shiny black and embedded in a sticky bright yellow fluid. The seeds would stick to birds' beaks and be transferred from tree to tree.

150. (Above right) A close view of an open capsule of *Pittosporum cornifolium*.

151. (Left) A large-leaved, bright green crown of the shrub epiphyte puka (*Griselinia lucida*) at the base of a northern rata crown. Mangamuka Bridge, Northland.

145

At intervals slender, horizontal, girdling roots arise from the main root and often extend around the trunk of the supporting tree several times to form an anchoring network.[153] The large, shiny, handsome leaves of the puka are distinctive in being very lopsided at the base—of the two parts of the leaf on each side of the midrib one is always shorter than the other.

Broadleaf, a smaller-leaved relative of the puka, usually establishes on the ground. This species generally grows at higher altitudes than puka in the north of the country, descending to the lowlands in the south. It too may grow as an epiphyte in wet forests, although its descending root is not grooved. The mountain five-finger may also be epiphytic at times with roots descending to the ground. These two species usually grow as epiphytes beyond the altitudinal and latitudinal ranges of nest epiphytes and so would establish on mossy branches.

The commonest large tree that begins as an epiphyte is the northern rata, found throughout most North Island conifer broadleaf forests and in the north-west of the South Island. In this case the young plants do not appear to establish in the humus below nest epiphytes, but in bark crevices on branches of tall trees. The roots of young plants of northern rata often have conspicuous tuber-like swellings which may function for water storage. As with the puka, eventually a root or roots grow down to the ground usually on the shady side of the supporting trunk. When the ground is reached, after several decades, the main descending root and branches from it enlarge greatly and give rise to horizontal girdling roots that encircle the trunk and eventually become quite woody themselves.[157] For probably a century or more the rata root system branches and enlarges with some of the roots fusing together to enclose the trunk of the supporting tree in a sort of cage. Eventually the supporting tree dies[158] and rots away, leaving the northern rata as a tall tree standing on a contorted pseudotrunk of coalesced scaly roots.[159,160]

As with its unrelated counterparts in the tropics, the northern rata is sometimes referred to as a 'strangler', with the implication that the supporting tree is compressed and killed within the root cage of the rata. Popular writers often dramatise this idea, describing the northern rata variously as a 'predatory gangster', 'forest bandit', or 'notorious strangler' that 'crushes', 'smothers', 'stifles' or 'squeezes' the supporting tree in an 'iron', 'deadly' or 'fatal' embrace! However, it has been pointed out that a northern rata would not usually establish on a tree that was not already mature with a well-lit crown. So by the time the rata itself reaches maturity the supporting tree would have died of old age, hastened no doubt by overshading by the crown of the epiphyte and the competition of its roots.

Southern rata is usually terrestrial, but in wetter parts of the South Island it may also grow as an epiphyte.

The relatively recently discovered and described *Metrosideros bartlettii* from North Cape is also frequently epiphytic, sometimes on puriri, but small plants have also been observed on tree fern trunks. Here too the descending roots eventually form pseudotrunks.[161]

152. (Page 146) Descending young roots of a puka pressed closely to a rimu trunk. (Photo: M.D. King)

153. (Page 147) A large descending root of a puka firmly attached with many girdling roots to the supporting kohekohe trunk. Battlehill Reserve, north of Wellington.

154. (Page 148) Nearer the ground the descending puka root often does not have girdling roots. Note the distinctive longitudinal grooves in the root. The supporting tree is a pukatea (*Laurelia novae-zelandiae*). Otari Native Botanic Garden, Wellington.

155. (Page 149) An unusual corkscrew root of a puka. This appears to have started life on a tree that died. The puka fell against a karaka and the puka is now anchored to it with a network of roots. Wi Perata Reserve, Waikanae.

Epiphytes on tree ferns

Tree ferns do not branch so the only surface available for epiphytes is that of the trunk. The trunks are not woody and do not have bark, but instead are covered by: the projecting bases of former leaves;[80] or hard scars where leaves were formerly attached; or masses of wiry roots that grow out from the stem between the leaf bases.[80,81] It has been observed that in some of the tree fern species the old dead fronds at the base of the crown persist indefinitely and, as they hang downwards, form a skirt around much of the trunk. It has been proposed that these skirts serve the purpose of preventing the establishment of epiphytes and vines on the trunks.

In *Cyathea smithii* of cooler forests only the stalks and midribs of the fronds persist like a cluster of hanging sticks.[162] In wheki-ponga the entire fronds persist to form an extremely thick skirt around trunks[163] that are massively enlarged with a layer of wiry roots. On the Chatham Islands, the only epiphyte I noted on such wheki-ponga was the fern *Rumohra adiantiformis* growing at the vulnerable junction between living and old fronds. However, it is significant that a group of wheki-ponga, exposed by forest clearance and with their skirts removed by wind, now support a number of woody epiphytes, some with roots descending to the ground. Among the latter are the tree *Dracophyllum* (*D. arboreum*) and the Chatham Island lancewood (*Pseudopanax chathamicus*). The spongy root layer of wheki-ponga trunks seems an ideal site for seedlings in the absence of skirts. When young the mamaku also has an untidy skirt of dead leaves, but in tall older specimens the leaves separate cleanly leaving an armour of hard leaf scars that prevent the establishment of seedlings.

156. (Above) Even the most specialised epiphytes can sometimes grow on the ground, usually in well-lit rocky sites. Here a puka is growing near the sea at Napenape, North Canterbury.

157. (Page 152) Northern rata (*Metrosideros robusta*) epiphytic on a rimu. The main descending root on the right is firmly anchored by now woody girdling roots. Kaitoke Regional Park, Wellington.

158. (Page 153) A northern rata now free-standing on its coalescing root system. The dead and now decaying original supporting tree can still be discerned within the rata roots. Tunnel Reserve, Te Marua, Upper Hutt.

159. (Page 155) A 'pseudotrunk' of northern rata roots. The nest epiphyte is *Astelia solandri*. Wainui Waterworks Reserve, Wellington.

160. (Page 154) A tree combination. At the top the spreading crown of a northern rata, at its base on the left the smaller rounded crown of a puka, then a mass of nest epiphytes and finally, lower-most, the drooping foliage of the rimu that supports them all. Kaitoke Regional Park, Upper Hutt.

161. (Above left) Pseudotrunk of roots of the recently described rata, *Metrosideros bartlettii*. Radar Bush near Cape Reinga. (Photo: John Dawson)

162. (Above right) An impressive grove of unusually tall *Cyathea smithii* tree ferns. In this species the stalks and midribs of the fronds persist like a skirt of sticks. Near Erua, central North Island.

163. (Left) A wheki-ponga tree fern (*Dicksonia fibrosa*) with a thick skirt of complete dead fronds. Pelorus River, northeast South Island.

164. (Opposite) The wheki tree fern (*Dicksonia squarrosa*), by means of underground rhizomes, often forms groves, particularly in swampy sites. Its dead leaves become very brittle and drop off to form an impressive litter on the forest floor.

In old and tall tree ferns with skirts, of course, the skirts do not reach the ground so epiphytes can establish below them.

The two common tree ferns that do not form skirts are ponga and wheki. In these the dead fronds break off readily, leaving projecting stalk bases that gradually rot to form a humic soil in the spaces between. As a consequence epiphytes, and indeed vines such as the climbing ratas and kiekie, are frequent on the trunks of these tree ferns. Of smaller epiphytes some mosses and several filmy ferns are largely restricted to these tree fern trunks. The larger fern *Rumohra adiantiformis* has already been mentioned. There is also a small form of *Lycopodium varium* and several species of the 'living fossil' genus *Tmesipteris*[165] which is considered to be one of the most primitive vascular plants. The several species have drooping, mostly unbranched stems with small leaves and unusual sporangia with two compartments.

165. (Page 158) A 'living fossil'. *Tmesipteris* is a genus considered to be descended from an ancient group of land plants. This is *T. tanneensis* epiphytic on the trunk of a ponga.

166. (Page 159) A ponga tree fern (*Cyathea dealbata*) with two epiphytic young five-fingers (*Pseudopanax arboreus*). Otari Native Botanic Garden, Wellington.

167. (Opposite) A fully mature five-finger on a ponga. The trunk of the five-finger extends up to the left and that of the ponga to the right. Below their junction is the root system of the five-finger growing through the zone of wiry roots of the tree fern. Otari Native Botanic Garden, Wellington.

168. (Left) Another example of a five-finger on a ponga, this time with one root ascending then descending. Note the descending kiekie roots at left background.

There are also a number of woody epiphytes that establish on tree fern trunks. Notable among these is the five-finger, whose seedlings establish below the crown of a tree fern.[166] The roots of the seedlings gradually grow down through the leaf bases and then through the denser mass of wiry roots near the base of the trunk until they reach the ground and begin to enlarge.[167,168] While this is going on the stem of the five-finger has developed into a trunk which carries the foliage above the crown of the tree fern. The related raukawa is also a frequent tree fern epiphyte.[169] This is the case too for the much larger tree kamahi. Here the seedlings establish nearer the ground,[170] so the kamahi eventually has a contorted base of roots below a number of spreading trunks. The tree ferns can sometimes be found still living at the centre of quite large trees,[171] but more often the tree ferns die and break off leaving a stump. In the far north the related towai and *Ackama rosifolia* of the same family also often establish on tree ferns.

The mistletoe parasites

These attach to the stems of trees and shrubs. Having leaves and stems with chlorophyll they are able to manufacture carbohydrate nutrients by photosynthesis and are only partial parasites. Their special absorbing organs (haustoria) at points of attachment penetrate into the living tissues of their hosts[174] and tap into the latter's supply of water and mineral nutrients drawn up from the ground.

Three of our mistletoes (*Korthalsella*) are only a few centimetres long and have very reduced scale-like leaves and tiny flowers. The stem segments between joints are cylindrical in one species and flattened and

169. (Page 162) Raukawa (*Pseudopanax edgerleyi*) epiphytic on a ponga. Tunnel Reserve, Te Marua, Upper Hutt.

170. (Page 163) Seedlings of kamahi (*Weinmannia racemosa*) established among the wiry roots of a ponga.

171. (Opposite) Kamahi trees established on a trunk of a tall *Cyathea smithii* tree fern. The large lower trunk belongs to a tree about 15 metres in height. The tree fern is about 10 metres high. Near Erua, central North Island.

172. (Above) The parasite *Korthalsella lindsayi* whose flattened stem segments mimic the leaves of its small-leaved host *Melicope simplex*. Carter's Bush, Wairarapa.

almost circular in the others.[172] The korthalsellas parasitise small-leaved shrubs, including manuka, mostly at forest margins and openings.

The other mistletoes, in a different family, are much larger with well developed leaves. They form rounded shrubs a metre or more in diameter attached to the branches and trunks of trees. In most of them stems radiate from the main point of attachment[173] and form secondary haustoria where they touch the host trunk. Two of these larger mistletoes have small green flowers (*Ileostylus, Tupeia*). They parasitise mostly smaller trees, although also totara, in conifer broadleaf forests. Three other species have larger, initially tubular flowers that are bright red[173] in two cases and yellow in the other (*Alepis, Peraxilla*). These mostly parasitise beech trees providing bright splashes of colour against the dark green foliage. Unfortunately, the depredations of possums have reduced these displays in more recent times. The mistletoes form sticky seeds some of which attach to the beaks of birds and are transported to other trees.

173. (Opposite) The scarlet mistletoe (*Peraxilla tetrapetala*) growing on the trunk of a mountain beech. Sucker-like haustoria can be seen here and there growing from the spreading stems of the parasite into the living tissues of the host. Lake Ohau, South Island.

174. (Above) A silver beech branch with attached mistletoe parasite cut through to show the invading suckers (*haustoria*) of the parasite.

9. Problems and prospects

In the preceding chapters we have explored the roles of plants in established forests similar to those that largely covered New Zealand in pre-human times. But even in those earlier times, before the most disruptive of all animals arrived, forest was destroyed from time to time. This happened locally from the collapse of old trees and by wind throw, or by landslides on our steep terrain triggered by heavy rain or earthquake; and more widely from fire following lightning strike in forests of drier regions and from the fallout from volcanic eruptions.

Following these disruptions the forest reconstitutes itself in various ways. In conifer broadleaf forest, smaller canopy gaps are soon occupied by a variety of quickly growing shrubs and small trees including tree ferns, and eventually taller trees establish below them and grow up to close the canopy. Where forest is totally destroyed over wider areas, particularly by fire, the barren landscape may first be covered by dense bracken fern, followed by the small-leaved shrubs to small trees, manuka and kanuka.[175] In the shelter of these, taller larger-leaved shrubs and small to larger trees grow up, such as five-finger, lancewood, karamu, kamahi and rewarewa. Later still conifers may establish, including kauri in the far north, rimu more widely, mostly on slopes, with totara, matai and kahikatea favouring alluvial sites. After several centuries shade-tolerant trees, including tawa and taraire, grow up below the conifers to form a closed canopy.

175. A stand of kanuka (*Kunzea ericoides*) that established following destruction of conifer broadleaf forest by fire. Under the mature kanaka canopy pioneer forest species have established—several species of tree fern, kamahi, five-finger and putaputaweta (*Carpodetus serratus*). Near Wainui Waterworks Reserve, Wellington.

"

The beech species require a fair amount of light to grow through to
maturity, although seedlings and saplings can establish on the forest
floor and then 'stand still' until more light is available. This is provided
when a canopy gap is formed, perhaps by the collapse of an old tree,
and the waiting beech seedlings quickly fill the gap. Where beech forest
is destroyed over larger areas beech seedlings may form a dense cover
on bare ground or grow up in the shelter of manuka or kanuka. The
rate of regeneration of beech depends on whether forest destruction
coincides with a good seed year. The beech species flower and set
abundant seed only at intervals of several years, usually following a
warm summer.

With the coming of the Maori the area of closed forest was reduced by
fire and the area of regenerating forest would have increased. The
European settlers greatly accelerated this process, although in this case
much of the land cleared of forest was sown to pasture and intensively
grazed, so preventing any regeneration. Nevertheless, regeneration is
taking place on steeper terrain unsuitable for farming where fire pen-
etrated and on other land that proved uneconomic and is no longer
farmed. So, in addition to the approximately 23 per cent of the country
still under native forest, there is also an additional area of regenerating
forest, which may eventually increase our native forest estate. Native
forest regeneration should certainly be encouraged for cultural and
scientific reasons, but also for the practical reasons of stabilising slopes
and reducing accelerated erosion.

However, preserving the surviving native forests and encouraging their
regeneration elsewhere would not be enough. As well as destroying

176. One of the giant
grasses known as toe-
toe (*Cortaderia fulvida*),
which may establish as
a pioneer on river
terraces, as here,
following forest
destruction. Near
Wainui Waterworks
Reserve, Wellington.

forests Maori and European settlers also introduced a number of
animals that have had a deleterious effect on still intact forests and on
the stages in their regeneration. In the case of the Maori this involved
only the kiore (rat), which eats and destroys the seeds of many native
trees. The European introduced a much wider range of animals that
became established in the wild, notably deer, goats and the Australian
possum. In earlier decades deer became abundant in native forests and
as a result the undergrowth was trampled and palatable species, including young plants of dominant trees, were eaten out. A profitable
market for venison in Europe led to intensive hunting of deer, most
effectively by helicopter, for both export and farming. This has led to a
considerable reduction in deer numbers and some forest recovery.
However, markets could change, deer hunting could diminish and the
deer population in forests could explode again.

The possum is currently a greater problem. There are many millions of
them in our forests and even in our streets and gardens. Being able to
climb, they attack tree crowns and eat leaves, flowers and fruits. Five-
finger,[177] the ratas[178] and mistletoes are particularly vulnerable and
often die as a result of defoliation. At one time possums were exten-
sively trapped for their fur, but this trade has largely collapsed with the
falling from favour of natural fur for clothing. Locally, possum
populations can be reduced by trapping and poisoning, or even eliminated on islands such as Kapiti, but this is not a practical, countrywide
solution.

To a lesser extent introduced plants have impinged on native forests. In
the larger forest stands, particularly those remote from settlements, this

177. Five-finger
(*Pseudopanax arboreus*)
that has been attacked
by a possum.

is not a problem, but in smaller remnants in settled areas such adventives as *Tradescantia* may cover the forest floor and Old Man's Beard (*Clematis vitalba*) smother tree crowns. On the other hand, at least one adventive plant, gorse, plays a beneficial role. In such places as Wellington's hills it establishes following fire and provides a very effective nurse crop for the pioneers of the native forests.

Finally, is the future for our remaining native forests hopeful? More hopeful probably than in colonial times. One can imagine the relatively few people at that time who valued the forests for aesthetic and intellectual reasons believing that it was only a matter of time before they completely disappeared. The prevailing view then was that the forests were obstacles to be removed or a resource to be exploited. This view has not entirely disappeared, but more and more people, we believe, are coming to regard our forests as a heritage from the distant past that should be preserved.

We hope that this book will interest and inform the converted, convert others, and so strengthen the case for conservation of our forest legacy from the ancient continent of Gondwana.

178. A dead northern rata (*Metrosideros robusta*) that probably died as a result of defoliation by possums. Note that the nest epiphytes are still alive, proving that they are not parasites. There is also a close growth of the white climbing rata (*Metrosideros perforata*). Orongorongo Valley, near Wellington.

Acknowledgements

We are grateful to Merv King, Peter Newsome, Craig Potton and Barry Sneddon for providing one photograph each and to the last also for his ongoing encouragement and support.

We thank Mary Murray for her patience with the typing and our editor, Fergus Barrowman, for his enthusiasm for the book and his skill in designing it.

We are especially grateful to the Lottery Grants Board for a generous grant to assist with publication costs.

Useful References

Brownsey, P.J. and Smith-Dodsworth, J.C. *New Zealand Ferns and Allied Plants.* David Bateman, 1989.

Cockayne, L. *New Zealand Plants and their Story* 4th ed. Government Printer, 1967.

Dawson, J.W. *Forest Vines to Snow Tussocks: The Story of New Zealand Plants.* Victoria University Press, Wellington, 1988.

— *The Ancient Forests of New Zealand* (video). Learning Media, Ministry of Education, 1991.

Eagle, Audrey. *Eagle's Trees and Shrubs of New Zealand.* Collins, Auckland, 1975. Second Series, 1982.

Enting, Brian and Dawson, John. *Seasons in the Forest. A New Zealand Photographer's Year.* Random Century, Auckland, 1990.

Poole, A.L. and Adams, N.M. *Trees and Shrubs of New Zealand,* Rev. ed. D.S.I.R. Publishing, 1990.

Salmon, J.T. *The Native Trees of New Zealand.* A.H. and A.W. Reed, Wellington, 1980.

Wardle, P. *Vegetation of New Zealand.* Cambridge University Press, 1991.

Glossary of Common names

Where a name involves an adjective
and a noun it is listed under the noun.

Beech, black *Nothofagus solandri* var.
 cliffortioides
Beech, hard *Nothofagus truncata*
Beech, mountain *Nothofagus solandri*
 var. *cliffortioides*
Beech, red *Nothofagus fusca*
Beech, silver *Nothofagus menziesii*
Beech, southern *Nothofagus*
Broadleaf *Griselinia littoralis*
Cabbage tree, lowland *Cordyline*
 australis
Cabbage tree, mountain *Cordyline*
 indivisa
Cedar, mountain *Libocedrus bidwillii*
Fern, crape *Leptopteris superba*
Fern, crown *Blechnum discolor*
Fern, hen and chicken *Asplenium*
 bulbiferum
Fern, kidney *Trichomanes reniforme*
Fern, king *Marattia salicina*
Fern, Prince of Wales Feathers
 Leptopteris superba
Fern, umbrella *Sticherus*
 cunninghamii
Ferns, filmy *Hymenophyllum,*
 Trichomanes
Five-finger, mountain *Pseudopanax*
 colensoi
Five-finger *Pseudopanax arboreus*
Fuchsia, tree *Fuchsia excorticata*
Grass, bush rice *Microlaena avenacea*
Grass, kauri *Astelia trinervia*
Hangehange *Geniostoma rupestre*
Haumakaroa *Pseudopanax simplex*
Heketara *Olearia rani*
Hinau *Elaeocarpus dentatus*
Horopito *Pseudowintera axillaris*
Horopito, mountain *Pseudowintera*
 colorata
Jasmine, native *Parsonsia heterophylla*
Kahikatea *Dacrycarpus dacrydioides*
Kaikawaka *Libocedrus bidwillii*
Kaikomako *Pennantia corymbosa*
Kanono *Coprosma grandifolia*
Kamahi *Weinmannia racemosa*
Kanuka *Kunzea ericoides*
Karaka *Corynocarpus laevigatus*
Karamu *Coprosma robusta*
Kauri *Agathis australis*

Kawaka *Libocedrus plumosa*
Kawakawa *Macropiper excelsum*
Kidney fern *Trichomanes reniforme*
Kiekie *Freycinetia banksii*
Kohekohe *Dysoxylum spectabile*
Kohuhu *Pittosporum tenuifolium*
Kotukutuku *Fuchsia excorticata*
Kowhai *Sophora* spp.
Lacebark *Hoheria* spp.
Lancewood *Pseudopanax crassifolius*
Lawyer, bush *Rubus cissoides*
Lawyer, swamp *Rubus australis*
Lemonwood *Pittosporum eugenioides*
Mahoe *Melicytus ramiflorus*
Maire *Nestegis* spp.
Maire, black *Nestegis cunninghamii*
Mairehau *Phebalium nudum*
Makomako *Aristotelia serrata*
Mamaku *Cyathea medullaris*
Mangeao *Litsea calicaris*
Mangemange *Lygodium articulatum*
Mangrove *Avicennia resinifera*
Manuka *Leptospermum scoparium*
Matai *Prumnopitys taxifolia*
Mingimingi *Cyathodes* spp.
Miro *Prumnopitys ferrugineus*
Mistletoe, scarlet *Peraxilla tetrapetala*
Monoao *Halocarpus kirkii*
Neinei *Dracophyllum latifolium*
Ngaio *Myoporum laetum*
Nikau palm *Rhopalostylis sapida*
Old Man's Beard *Clematis vitalba*
Orchids, green hood *Pterostylis* spp.
Orchids, spider *Corybas* spp.
Parataniwha *Elatostema rugosa*
Passion vine, native *Passiflora*
 tetrandra
Pate *Schefflera digitata*
Pigeonwood *Hedycarya arborea*
Pine, bog *Halocarpus bidwillii*
Pine, celery *Phyllocladus* spp.
Pine, silver *Lagarostrobus colensoi*
Pine, yellow-silver *Lepidothamnus*
 intermedius
Pohutukawa *Metrosideros excelsa*
Pokaka *Elaeocarpus hookerianus*
Ponga *Cyathea dealbata*
Pua reinga *Dactylanthus taylori*
Puka *Griselinia lucida*
Pukatea *Laurelia novae-zelandiae*
Puriri *Vitex lucens*
Putaputaweta *Carpodetus serratus*

Index

Common names are used where possible. Ferns and their allies and fungi are indexed separately from seed plants. Plants with special life styles are grouped under relevant headings. Figures in italics refer to illustrations.